Desertification: Financial Support for the Biosphere

Desertification:
Financial Support for the Biosphere

Yusuf J. Ahmad
Mohamed Kassas

Sponsored by the United Nations Environment Programme

HODDER AND STOUGHTON
LONDON SYDNEY AUCKLAND TORONTO

ISBN 0 340 41044 2

First published 1987
Copyright © 1987 by the United Nations Environment Programme
All rights reserved. No part of this publication may be
reproduced or transmitted in any form or by any means,
electronic or mechanical, including photocopy, recording,
or any information storage and retrieval system, without
permission in writing from the publisher.

Photoset by Rowland Phototypesetting Ltd
Bury St Edmunds, Suffolk
Printed in Great Britain for Hodder and Stoughton Educational,
a division of Hodder and Stoughton Ltd,
Mill Road, Dunton Green, Sevenoaks, Kent
by Richard Clay Ltd, Bungay, Suffolk

Contents

Abbreviations — vi
Preface — viii
Introduction — xi
Editor's Notes and Acknowledgments — xix

PART I   Study of Additional Measures and Means of Financing the Implementation of the Plan of Action to Combat Desertification – 1978 — 1

PART II Study on Financing the United Nations Plan of Action to Combat Desertification – 1980 — 25

PART III Feasibility Studies on and Detailed Modalities for Financing the Plan of Action to Combat Desertification – 1981 — 103

Index — 183

Abbreviations

Acronym	Meaning
ACC	Administrative Committee on Co-ordination
CF	Common Fund
CGIAR	Consultative Group on International Agricultural Research
DC/PAC	Desertification Control Programme Activity Centre
DECARP	Desert Encroachment Control Rehabilitation Programme
DESCON	Consultative Group for Desertification Control
DFI	Development Financing Institutions
EEZ	Exclusive Economic Zones
FAO	Food and Agriculture Organization
GDP	Gross Domestic Product
GEMS	Global Environment Monitoring System
GNP	Gross National Product
ICA	International Commodity Agreement(s)
ICARDA	International Center for Agricultural Research in the Dry Areas
ICASALS	International Centre for Arid and Semi-Arid Land Studies
ICB	International Commodity Body
ICRISAT	International Crops Research Institute for the Semi-Arid Tropics
IDA	International Development Association
IDB	International Development Bank
IDRC	International Development Research Center
IFAD	International Fund for Agricultural Development
IFC	International Finance Corporation
IFPRI	International Food Policy Research Institute
IFRB	International Frequency Registration Board
ILCA	International Livestock Center for Africa
ILO	International Labour Organization
IMF	International Monetary Fund
INTELSAT	International Telecommunications Satellite Organization
ITU	International Telecommunications Union
LDC	Least Developed Countries

MFN	Most Favoured Nation
MFR	Maximum Financial Requirements
MIT	Massachusetts Institute of Technology
NGO	Non-Governmental Organization
ODA	Official Development Assistance
OECD	Organization for Economic Co-operation and Development
OPEC	Organization of Petroleum Exporting Countries
PACD	Plan of Action to Combat Desertification
SAREC	Swedish Agency for Research Co-operation with Developing Countries
SDR	Special Drawing Right(s)
SITC	Standard International Trade Classification
UNCOD	United Nations Conference on Desertification
UNCTAD	United Nations Conference on Trade and Development
UNDP	United Nations Development Programme
UNEP	United Nations Environment Programme
UNESCO	United Nations Educational, Scientific and Cultural Organization
UNITAR	United Nations Institute for Training and Research
UNSO	United Nations Sudano-Sahelian Office
UNU	United Nations University
WMO	World Meteorological Organization

Preface

The United Nations Conference on Desertification (UNCOD, 1977) showed that desertification was one of the major environmental problems of the world. Its cost in human, social and economic terms is extremely high. The Conference adopted a Plan of Action to Combat Desertification (PACD) that set as its objective the halting of desertification by the year 2000 and enumerated actions required to achieve that objective. In 1984 I reported to the Governing Council of United Nations Environment Programme (UNEP), and through it to the world, on a general assessment of the status and trend of desertification. The main findings can be summarized as follows:

- the scale and urgency of desertification as presented to UNCOD (1977) and addressed by the PACD are confirmed; desertification has continued to spread and intensify despite efforts undertaken since 1977;
- land irretrievably lost through desertification continues at about six million hectares annually, and land reduced to zero or negative economic productivity is more than 20 million hectares annually;
- areas of productive land affected by at least moderate desertification include 3100 million hectares of rangeland, 335 million hectares of rainfed croplands and 40 million hectares of irrigated land;
- rural populations in areas severely affected by desertification number 135 million (1984), compared with 57 million in 1977.

A 20-year world-wide programme to arrest further desertification would require about $4.5 billion a year; developing countries in need of financial assistance would require $2.4 billion of this ($48 billion in the 20 years). Clearly these figures are beyond present levels of donor assistance (1980 estimate: $600 million annually for anti-desertification related activities). Additionality of resources is thus a basic element in the combat against desertification. But the long-term nature and magnitude of the problem is such that resources will be needed which are not only additional but sustained, predictable and assured.

The cost of arresting desertification should not be considered, as it usually is, as prohibitively high. On the contrary, it is common sense to establish means of raising such resources when we know that the process of desertification makes a significant contribution to the degradation of life-sustaining biogeochemical cycles, spread abject poverty and loss of human life and that the losses in productive capacity because of this amount to nearly $16 billion per year.

It was in recognition of these compelling reasons that I had raised during UNCOD (1977), in my capacity as the Secretary-General of the Conference, the issue of the need for 'automaticity' in the international assistance for the implementation of the PACD. This was the first time that the question of 'automaticity' was presented in a UN forum and after extensive debate on what was considered a path-breaking and innovative principle, the PACD adopted by the Conference requested a study of additional measures and means of financing including 'fiscal measures entailing automaticity'.

Since then, a series of three studies carried out by high-level specialists in the international financing of projects and programmes have considered a variety of possible fund-raising modalities entailing automaticity. These have included revenues from the exploitation of international commons, taxation of defence expenditures and arms transfers, a development link with special drawing rights (SDR), the taxation of international trade, a Common Fund for Commodities (Second Account) and yet others. Proposals and ideas examined are diverse: some have been formally mooted within the UN system, others are largely speculative, some are close to realization or already initiated, others are clearly long-term. In all instances, the studies have attempted to evaluate schemes from the point of view of their economic efficiency and practicability.

One point which emerges and which deserves to be emphasized is that some of the means of financing identified could have the potential of enlarging the totality of resources available to meet increasingly critical environmental and natural resource concerns of developing countries without placing an undue burden on the national budgets of developed countries.

A particularly interesting proposal to emerge from the studies relates to the creation of an independent operating financial corporation for the financing of desertification projects. The following features of the proposal are noteworthy: it is intended to operate with a predictable flow of interest-free loans contributed by the majority of UN member countries to be replenished after intervals of seven years; the corporation is expected to operate at the extreme end of the project feasibility scale and finance those long-term desertification projects which, although possessing definable economic and social viability, will produce returns which cannot be quantified at the time of project preparation, and the ultimate borrower of interest-free money will be offered loans for periods of between 40 and 45 years.

The three studies were duly presented to the General Assembly. The long debate in the General Assembly over the past several years during which no conclusive recommendations could be reached has shown that the practical problems involved, complex as they are, are not the real obstacle towards meeting, in an orderly and effective manner, the needs of development and environmental protection; the real obstacle is the question of political will. Is the international community prepared to recognize its undeniable responsibility and embark upon measures that offer a clear potential for the mobilization of adequate resources, or continue with its present neglect of problems which are becoming more costly and more compounded with the passage of time? The three studies have been edited and presented in this book so that the valuable work done is not lost and that there is wider dissemination of ideas and concepts which may take a longer time for their realization but action upon which should commence now without further delay.

Nairobi, August 1986

Mostafa K. Tolba
Executive Director
United Nations Environment Programme

Introduction

A Study on *New Means of Financing International Needs* was published by the Brookings Institution in Washington, DC, in 1978, jointly authored by Eleanor B. Steinberg and Joseph A. Yager with the co-operation of Gerard M. Brannon.[1]

The authors of the study argued that voluntary contributions by national governments to the United Nations system and funds borrowed in capital markets by the World Bank and other international financial institutions could not be counted upon to meet growing international financial requirements. They were of the view that financing methods would be needed to cope with the world's environmental problems and to help the developing countries raise the living standards of their people. The study concluded that a new, more powerful international revenue system could be created within the present international political order, that is, in a world of sovereign nation states. They recognized, however, that governments may not yet be ready to act jointly to raise large amounts for international purposes.

The situation has remained essentially the same during the last eight years except in one respect. The UN Conference on Desertification, which was held a year earlier (1977) than the publication of the study, had proposed a Plan of Action to Combat Desertification (PACD). The PACD envisaged that its implementation would be 'carried out by governments through their national institutions', and hence recommended the creation and development of national bodies and machineries that would be capable of implementing national plans and the provision – from national sources or with aid support – of the human and material resources needed for the effective operation of the national machinery. National programmes have sought to reach three principal objectives: (1) arrest desertification processes, (2) establish ecologically sound land-use practices that ensure sustained productivity, and (3) provide for social and economic advancement of the communities concerned.

But action at the national level remained – in general – far from being effective. The countries that are worst affected are the Third World countries in tropical drylands and adjoining sub-humid territories. These countries are disadvantaged climatically (recurrent droughts) and economically (low-income countries) and are often beset by political unrest. Moreover, desertification control projects do not seem capable of competing for the limited financial resources that are available; their rates of economic return are often lower, and their gestation periods are often longer, when compared with other marketable projects. With these constraints, countries appear unable to accord priority to desertification control programmes within their plans for investment of the all-too-scarce cash and capabilities.

An underlying theme of the financial requirements for the implementation of the PACD is derived, *inter alia*, from General Assembly resolution 3362 (5-VIII) adopted on 19 September 1975, which reads in part:

> 'Concessional financial resources to developing countries need to be increased substantially, their terms and conditions ameliorated and their flow made predictable, continuous and increasingly assessed so as to facilitate the implementation by developing countries of long-term programmes for economic and social development.'

If long-term programmes for economic and social development required concessional financial resources on predictable, continuous and increasingly assessed terms, major environmental threats such as desertification, would clearly deserve similar support. It was estimated by UNCOD that a 20-year programme to combat the encroachment of deserts would cost approximately $4.5 thousand millions a year.[2]

Available international means for financing such a programme are becoming increasingly scarce, not only for UNEP which is charged with the management and supervision of the implementation of the PACD but also for other UN Agencies and bodies and multilateral financing institutions on the one hand and bilateral aid agencies on the other. The contributions to the Environment Fund do not enable UNEP to plan to meet even a fraction of the totality of funds required. In so far as the UN specialized agencies are concerned the contributions received by them are all earmarked funds and cannot be directed in large amounts to global environmental concerns such as desertification. Multilateral financing institutions, including the World Bank, the International Monetary Fund (IMF) and the regional development banks, are bound by mandates for which they were created and cannot channel their resources substantially into programmes of their choice, however deserving. Thus, the capacity of the International Development Association (IDA) of the World Bank to offer loans on highly concessional terms, which are essential to finance a long-term programme such as the PACD, depends upon the willingness of the major donors to replenish its level. Such willingness can only be characterized as minimal in recent years.

The combat of desertification cannot be carried out by borrowings from the capital markets unless and until the concessional terms of the market are made acceptable by blending with contributions from sources such as IDA without charge. This leaves the possibility of funding with bilateral aid agencies and such organizations as the Organisation of Petroleum Exporting Countries (OPEC), but such fundings are not normally linked to the needs of a structured global plan of action designed to meet compelling needs outside the network of existing or expedient political and other external considerations.

Actions of combatting desertification are inseparable from actions of resource development and management in arid and semi-arid lands. Schemes that aim at checking land-degradation in pasturelands, rainfed farmlands and irrigated agricultural lands; at sand-dune stabilization; at establishing large-scale green belts; at introduction of soil and water conservation systems in resource management; or at reclaiming new areas of arid and semi-arid lands, are apt to be costly. Projects involving irrigation schemes are particularly expensive. Such projects are generally non-competitive in terms of market values, especially when compared with prevalent rates of interest. Investments in land-reclamation projects commonly do not pay well financially, but their social and humanitarian values as means of ensuring food security and participation in production are immense.

This situation represents compelling reality that exists not only for the PACD but for all large environmental programmes which are inherently not self-financing or where a threshold level of capital is needed for effective action – such programmes as the World Soils Policy, the Global Plan of Action for Marine Mammals, GEMS, environmental health programmes to deal effectively with malaria, schistosomiasis, cotton pests and others, pollution of the oceans and of the atmosphere and yet others.

It was in this context that the UN General Assembly decided in 1977 to invite the Governing Council of UNEP to have prepared, by a small group of high level specialists in the international financing of projects and programmes, a study of additional measures and means of financing the implementation of the Plan of Action to Combat Desertification as recommended by UNCOD.

The PACD contains objectives, guiding principles and a set of 28 interrelated recommendations for combatting desertification. Its immediate goals are to arrest and, where possible to reverse, the process of desertification and reclaim desertified land. The accomplishment of these objectives should, in economic and humanitarian terms, increase agricultural production and well-being of the people living in areas subject to desertification, increase their income and give them greater security against drought and other adversities.

An essential element in the implementation of the Plan of Action was seen to lie in the mobilization of sufficient resources to finance the capital investment, research, training and other programmes required.

The study which was eventually prepared by the group of high-level experts, convened by the Executive Director of UNEP, at UNESCO Headquarters in Paris on 13–16 March 1978, also took into account the other provisions of the General Assembly resolution on the question as well as the detailed recommendations contained in the Plan of Action, paragraph 104 (e) of which reads as follows:

'(e) *Additional measures*
'The General Assembly should be invited to request the Governing Council of UNEP to have prepared, by a small group of high-level specialists in international financing of projects and programmes, a study of additional measures and means of financing for the implementation of the Plan of Action as adopted by the Conference, such as funds-in-trust, fiscal measures entailing automaticity, and an international fund, and to submit a final report on the subject of additional measures of financing to the General Assembly at its thirty-third session, through the Economic and Social Council.'

In its report, the Group pointed out that it had reviewed a number of specific ways in which additional resources for financing the implementation of the Plan of Action could be generated, including some which involved automaticity. A range of possibilities that appeared to warrant careful consideration were included in the study. Some of the measures reviewed showed promise as immediate means of mobilizing funds; others were novel and could be positively considered by the international community.

This study constitutes the substantive basis of Part I of the present book.

The study was duly considered by the General Assembly of the United Nations. The General Assembly requested the Secretary General to solicit the views of Member Governments on the study and to report the results back to the General Assembly.

In 1979 the General Assembly requested 'a complete inventory of relevant ideas and proposals put forward in the United Nations system of possible new ways and means to finance programmes of multilateral organizations at the world level, additional to regular assessed budgets and conventional extrabudgetary resources'.

A group of high-level experts was convened in Geneva on 21–23 July 1980 to respond to the request of the General Assembly. The inventory presented in the study prepared by the group is the first of its kind, and the most comprehensive and up-to-date at the time it was completed.

The inventory showed that, although the United Nations Conference on Desertification was the first United Nations forum in which the subjects of international taxation and automaticity were discussed, ideas for generating international revenues on automatic bases have been of long standing within the United Nations. The inventory described proposals for international taxation; revenues from the use of the 'international commons'; income from military taxes and savings from disarmament; monetary measures such as a SDR-development link and a link between gold sales and development; commodity price stabilization mechanisms and orderly commodities development; and other miscellaneous proposals. The inventory also dealt with proposals concerning the related considerations of the mechanism of the mobilization of resources, burden sharing and the allocation of the proceeds. Finally, it contained a brief evaluation of the proposals presented.

The evaluation of the proposals contained in the inventory may broadly be divided into three categories:

(a) measures involving automaticity which have already been carried out and shown to be feasible, have been approved, or are nearing approval, e.g. the establishment of a trust fund from International Monetary Fund gold sales for instance to developing countries; the agreement to establish a Common Fund with two 'windows'; and various proposals made at the Third United Nations Conference on the Law of the Sea to use for international purposes revenues from the deep ocean bed and parts of continental shelves more than 200 miles from shore;
(b) proposals which, at least *prima facie*, appear feasible and merit detailed action oriented feasibility studies, e.g. international taxation, the SDR-development link and 'parking' fees for satellites in geosynchronous orbit; and
(c) proposals, the implementation of which does not at the moment appear on the horizon: some of these proposals may be aimed too far into the future or at present seem impractical; others deserve further detailed studies and concerted efforts by the international community to advance their eventual development, e.g. disarmament and development, and military taxation, even if it is unrealistic to expect agreement on the proposals at the moment.

The study noted that in the final analysis the prospects for the adoption of the proposals described in the inventory depended not so much on technical feasibility as on the political will of the Member States to whom the study was addressed.

This second study constitutes Part II of the present book.

In 1980, the General Assembly considered the study prepared by the experts and requested the Secretary General:

'(a) To prepare, in consultation with the United Nations Environment Programme and with the assistance of similar group of experts on the subjects concerned, to be convened by the Executive Director of the Programme:
  '(i) Feasibility studies and concrete recommendations for the

implementation of the additional means of financing deemed practicable by the Secretary General including those providing for a predictable flow of funds;

'(ii) The detailed modalities of obtaining resources on a concessionary basis;

'(iii) A full feasibility study and working plan for the establishment of an independent operational financial corporation for the financing of desertification projects;

'(b) To report on the results of the above-mentioned studies to the General Assembly at its Thirty-Sixth Session.'

The study, which was prepared by a group of high-level specialists in international financing who met in Geneva on 20–24 July 1981, was presented to the General Assembly. The study is contained in Part III, and it includes the feasibility study for the establishment of an international financial corporation for financing of desertification projects.

The General Assembly has since requested comments of the member Governments on the Report of the Secretary General and an indication as to whether the recommendations are acceptable. At the Fortieth Session of the General Assembly in 1985, the third report on the response of the Member Governments was submitted. The Report stated that 'there seems to be no general support for the series of measures for providing additional resources needed for financing the Plan of Action to Combat Desertification as outlined in the three expert studies. The General Assembly may, however, consider that the innovative approaches and modalities embodied in the three expert studies deserve further consideration. A small group with a more closely focused perception may be better able to perform this task. The Consultative Group for Desertification Control, established by the General Assembly in resolution 32/172 of 19 December 1977, could be asked to do so with a view to the creation of practical and effective means for the early financing of the implementation of the Plan of Action, with particular emphasis on the feasibility of establishing an international financial corporation for that purpose.'

It will be recognized that the different studies prepared by experts and reports submitted to the General Assembly are part of a continuum: a continuing search for an acceptable and viable solution to the problem of financing the Plan of Action including proposals for practicable sources of funding which originated during UNCOD and were later discussed by the General Assembly at its thirty-second and almost every subsequent session.

No solutions are yet in sight although the present crisis in Africa indicates that we are reaching a point when synergistic, and perhaps irreversible, forces are coming into play and the 'Time Bomb of Desertification' is nearing its point of explosion.

INTRODUCTION xvii

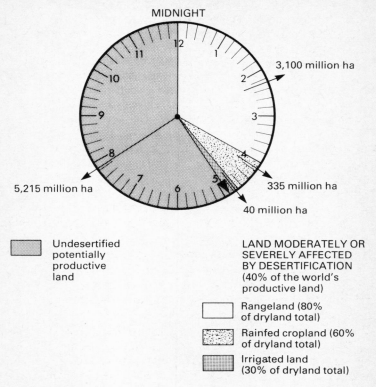

The Time Bomb of Desertification

The clock is ticking, an environmental time bomb. It is now 4:48 p.m. At the rate of 27 million hectares lost a year to the desert or to zero economic productivity, in a little less than 200 years at the current rate of desertification there will not be a single, fully productive hectare of land on earth. It will be the earth's midnight. We can expect the rate to increase as less land is exploited by more people. We can also be sure that a major socio-economic catastrophe of cataclysmic proportions would engulf the earth long before the last hectare succumbed. We are already feeling the first tremors of that catastrophe.

In 1983 the United Nations Environment Programme carried out a new assessment of the global dimensions of desertification as compared to an earlier assessment that was submitted to UNCOD in 1977. The new assessment showed that desertification-prone areas were not confined to arid and semi-arid regions and that extensive stretches of sub-humid tropics were threatened. The total areas prone to desertification covered 4,500 million hectares (35% of the Earth's land surface), human population of these territories were more than 850 millions (C.20% of the world population). The area of lands that were at least moderately damaged (3,475 million hectares) included:

3100 million hectares of rangelands;
335 million hectares of rain-fed croplands;
40 million hectares of irrigated farmlands.

The 1983 assessment confirmed the 1977 estimate that the world was losing on average an estimated 6 million hectares of productive land each year, desertified. It added that each year some 21 million hectares of productive land were being reduced to a state of zero or negative economic returns.

The Editors would like to emphasize that in the three reports covered in this presentation the political reality of today's climate of recession, growing unemployment, financial uncertainty, exchange rate fluctuations and, above all, the reluctance of many of the principal donors to allocate increased funds for international programmes was not ignored or minimized. But the experts and specialists were working on the hypothesis that enlightened self-interest would indicate that major environmental hazards, such as desertification, affected, through multiplier effects, the interests of all countries. Secondly, they were attempting to identify an armoury of weapons that may be employed when the political will emerges so that new modalities or sources of funding are not excluded from consideration simply because their early adoption, in the present climate of opinion, appears unlikely. Thirdly, the experts had made no *a priori* assumptions regarding the ushering in of a World Government with powers of international taxation: the international revenue taxes and other levies examined are situated in the context of the existing international political order.

Nairobi

Mohamed Kassas
Yusuf J. Ahmad

Notes

1 See Eleanor B. Steinberg and Joseph A. Yager with Gerard M. Brannon, *New Means of Financing International Needs* (Washington, DC, Brookings Institution; 1978).
2 This figure applies only to the cost of halting the further spread of deserts and not to the cost of reclamation of land which has already undergone desertification over the years.

Editors' Notes and Acknowledgments

The present study consists essentially of three United Nations reports prepared by groups of high-level specialists in international financing convened by the Executive Director of the United Nations Environment Programme. The three groups met at the UNESCO Headquarters in Paris on 13–16 March 1978 and at the Palais des Nations in Geneva on 21–25 July 1980 and 20–24 July 1981. The reports were presented to the United Nations General Assembly in documents A/33/260 of 6 October 1978, A/35/396 annex of 17 September 1980 and A/36/141 of 1 October 1981 respectively.

The distinguished specialists who participated in the three expert groups are listed in annexes to each of the three studies. Some of them participated in one meeting only, others in more, but all of them, serving in their personal capacities, made substantive contributions in terms of identification and evaluation of innovative ideas and concepts of financing. The three studies constitute a noteworthy tribute to their dedication and effort.

The Executive Director of UNEP, Dr Mostafa K. Tolba, attended each of the three meetings and served at once as the co-ordinator of the entire programme and catalyst of innovative ideas. The three studies bear his personal mark through out.

Several UNEP staff members also serviced the groups at their different meetings with data, information and analytical work which proved useful in the final preparation of the studies. The contribution of Mr Reuben Mendez of UNSO to the work of the first two groups was valuable and substantive and the editors wish to express their particular gratitude to him.

Acknowledgements of debt are also due to research personnel who worked behind the scenes in collating, collecting and checking diverse financial modalities and their practicability.

The successive drafts of the book were processed by Ms Betty Kiru, Ms Gillian Mayers and Ms Jane Maina with great care and attention and to them we are very grateful.

Yusuf J. Ahmad
Mohamed Kassas

Nairobi, August 1986

PART I

Study of Additional Measures and Means of Financing the Implementation of the Plan of Action to Combat Desertification (1978)

*Original Text:
General Assembly Document:
A/33/260 of 6 October 1978*

I INTRODUCTION

The United Nations Conference on Desertification, which met at Nairobi from 29 August to 9 September 1977, reviewed a number of technical documents, including a world map showing areas threatened by desertification.[1] Certain premises were discussed and accepted as bases for the Plan of Action to Combat Desertification (PACD) adopted by the Conference. These include the following:

(a) Desertification is a process of environmental degradation by which productive land is made non-productive and desert-like. The total area of once productive land lost in recent decades is estimated at over 9 million square kilometres, the present annual rate of loss being of the order of 60,000 square kilometres. The significance of this figure is indicated by the fact that the world's present food-producing land area is about 13 million square kilometres.
(b) Desertification is a global environmental problem, which directly affects some 100 countries in the world and indirectly affects the remainder. Its seriousness is accentuated by the fact that food production needs to be substantially increased to provide for the growing population of the world.
(c) Desertification and, in particular, its acceleration in recent decades is essentially due to interactions between societal systems and fragile ecosystems. These include actions of man, particularly the misuse and over-use of land resources. There is no evidence that desertification is caused by large-scale climatic change.
(d) The effort to combat desertification should be integrated into national development plans and priorities. The shortage of resources (capital, knowledge and trained manpower) represents a major obstacle to action for halting desertification and reclaiming desertified lands. Desertification is thus particularly overwhelming in the least developed countries.

The Plan of Action, which was adopted by the Conference and subsequently approved by the General Assembly, contains objectives, guiding principles and a set of 28 interrelated recommendations for combating desertification. Its immediate goals are to arrest and, where possible, to reverse the process of desertification and reclaim desertified land. The accomplishment of these objectives should increase agricultural production and contribute to global efforts to bridge the world food gap, improve the well-being of the peoples living in areas subject to desertification, and give them greater security against drought and other adversities.

An essential element in the Implementation of the Plan of Action is the mobilization of sufficient resources to finance the capital investment, research, training and other programmes required. To deal with this problem the General Assembly, at its Thirty-Second Session, called for certain actions for financing the implementation of the Plan of Action. Paragraphs 7 to 13 of resolution 32/172 read as follows:

'*The General Assembly* . . .

'7. *Requests* the organs, organizations and other bodies of the United Nations system to support international action to combat desertification within the context of the Plan of Action;

'8. *Decides* to entrust the Governing Council and the Executive Director of the United Nations Environment Programme as well as the Environment Co-ordination Board [Functions of this Board are now with the UN Administrative Committee on Co-ordination (ACC) – Editors], with the responsibility of following up and co-ordinating the implementation of the Plan of Action, in accordance with recommendation 27 thereof,[2] and requests the Governing Council to report, through the Economic and Social Council, to the General Assembly at its Thirty-Third Session and every two years thereafter;

'9. *Calls upon* all countries, in particular developed countries, as well as multilateral financial institutions and non-governmental donors, to provide and increase their assistance to countries suffering from desertification, especially for the financing of their subregional and regional programmes and projects within appropriate consortium arrangements, such as those pertaining to the Sahel green belt, and urges developing countries to give due priority to desertification problems in their development assistance requests;

'10. *Authorizes* the Executive Director to convene immediately a consultative group, which would meet as and when required, comprising representatives from the organizations referred to in paragraph 7 above, such other organizations as might be required, donor countries, multilateral financial agencies as well as developing countries having a substantial interest in combating desertification, to assist in mobilizing resources for the activities undertaken within the framework of implementing the Plan of Action;

'11. *Endorses* in principle the creation of a special account within the United Nations for implementing the Plan of Action;

'12. *Requests* the Secretary-General to prepare and submit a study on the establishment and operation of such an account to the General Assembly at its Thirty-Third Session, through the Governing Council and the Economic and Social Council;

'13. *Invites* the Governing Council to have prepared, by a small group of high-level specialists in the international financing of projects and programmes, a study of additional measures and means of financing for the implementation of the Plan of Action, and to submit a final report on the subject of additional measures of financing to the General Assembly at its Thirty-Third Session, through the Economic and Social Council.'

II THE NATURE OF THE PROBLEM

More than one third of the earth's land area, including the natural deserts, is arid. Some 628 million people, or 14 per cent [1977 estimates by UNEP–UNCOD studies, in a UNEP *General Assessment of Progress in the Implementation of the PACD:* 1977–1984 (UNEP/GC.12.9); population affected is estimated as 850 million or about 20% of the world's population] of the world's population, live in vulnerable drylands bordering the world's deserts. Of these, between 50 and 78 million are immediately affected by decreased productivity caused by desertification of semi-arid and even sub-humid lands. The losses wrought by the process are manifested in various ways, the most serious being the loss and degradation of human life. In Africa, where the problem is most severe, some countries are located entirely in arid or semi-arid areas and will continue to suffer substantial loss of productive land to desertification. Sixteen of the 30 'least developed' of the developing countries are suffering from desertification.

Desertification is caused mainly by man wrestling to secure a livelihood under ecological conditions which are essentially fragile. Lacking alternatives, he is often forced to resort to land-use practices that are destructive to the productive ecosystem. It can thus safely be said that desertification is caused by misuse or over-use of land. Proper development in such desertified areas would halt and even reverse the process of desertification and hence contribute to the well-being of the affected peoples.

The urgency and magnitude of the problem are illustrated by the following data: on the southern fringes of the Sahara, 650,000 square kilometres (65 million hectares) of once productive land have become desert during the last 50 years. Throughout the world, about 60,000 square kilometres (6 million hectares) are being lost annually. These include 3.2 million hectares of rangeland, 2.5 million hectares of rain-fed farmland and 125,000 hectares of irrigated farmland. Areas of greater size are partly damaged and their productivity reduced through soil erosion, salinization, water logging and other forms of soil degradation and loss of fertility. Desertification is both a major environmental hazard and a serious obstacle to development.

Despite the magnitude of the problem, there was, prior to the United Nations Conference on Desertification, no focus within the international community on desertification and no international programme designed specifically to combat it. Various members of the United Nations family and other intergovernmental bodies and bilateral and multilateral programmes, however, have dealt and continue to deal with aspects of development in arid lands, elements of which are concerned with combating desertification.

Because anti-desertification measures, especially on non-irrigated land, often do not produce immediate returns that are quantifiable, and because the countries concerned often have limited debt-servicing capacities, anti-desertification programmes can seldom meet the criteria of additional financial institutions. Again, action for halting desertification and reclaiming desertified land requires long-term programmes for which assured flows of financial resources are indispensable. There is clearly a need for special financial arrangements for combating this major threat to human welfare.

The governments concerned and the secretariat of the United Nations Conference on Desertification initiated in 1977 a set of studies on the feasibility of transnational projects to combat desertification. These projects require financing, and their execution would involve implementation of portions of the Plan of Action. Projects have been formulated for the following:

(a) The development and management of rangelands and livestock in the Sudano-Sahelian regions (SOLAR)[3]

Pre-investment studies, including pilot experimental projects and development of infrastructure and national and regional institutional support machineries, would take five to seven years and would require about $130 million. Implementation of the programme, in the light of these studies, would encompass rational development of resources in seven countries and would require, apart from reorganization of government machineries and socio-political patterns, substantial capital resources.

(b) The establishment of a green belt across the five countries to the north of the Sahara[4]

This would be a transcontinental belt of interconnected national projects involving controlled land-use, soil and water conservation, sand-dune stabilization, shelter-belts, managed rangelands, etc. The total cost of this belt, which will combine action to halt desertification with ecologically sound land use and development of natural resources, will be in the order of $50 to $70 million per year for at least 10 years.

(c) Green belt in Africa south of the Sahara, extending from the Atlantic to the Red Sea[5]

The cost of establishing this belt is estimated to be about $1.5 billion. This belt extends across the territories of some of the least developed countries, and international support would be needed in order to establish it.

(d) A transnational programme for the survey and development of the major regional aquifers in north-east Africa and the Arabian Peninsula[6]

Inventories and pilot projects that need to be carried out during the first five to seven years would cost from $30 to $35 million and would provide guidelines for rational use and management of extensive ground-water resources in these arid regions. The cost of development of these aquifers, including agricultural, industrial and domestic use, is estimated on the order of $1 billion over a period of five years. In the region covered by this programme, there are countries with considerable oil resources and hence capital. Others lack these resources and will need international support.

(e) The regional monitoring of desertification processes and related natural resources in:

(i) four countries in south-west Asia;[7]
(ii) four countries in South America.[8]

These two projects, envisaged as pilot experimental ventures, cover only 4 to 5 per cent of the total area of land prone to desertification. A global network for monitoring desertification will be needed.

It should be emphasized that the projects described above are only examples, presented to the Conference on Desertification, of likely areas of regional co-operation in anti-desertification measures. They by no means reflect the whole range of possible measures – national, regional and international – which will have to be undertaken to halt and reverse the spread of deserts. The Sudan, for instance, has elaborated a national plan, the Desert Encroachment Control Rehabilitation Programme (DECARP),[9] comprising nine pilot projects. The costs are estimated to be in the region of $16 million. The government considers that these projects could be carried out over a three-year period in the first instance. If similar programmes of pilot projects are undertaken in other countries of the Sudano-Sahelian region, the total cost for an initial five years would be approximately $130 million.

The global cost of corrective measures to prevent continuing net losses of land through desertification is estimated at $400 million annually [estimates of cost based on a detailed inventory contained in

Part II], exclusive of national and regional costs of infrastructure and administrative and other programme-support machinery. This is the order of magnitude of the minimum amount of funds required to achieve and maintain 'zero desert growth'. The benefits which would accrue would be in the region of $1,300 million annually, representing a benefit-cost ratio of over 3:1. This, however, would only be 'standing still'. To reverse desertification through reclamation would require considerably greater expenditure. A target reducing the area of desertified land by 6 million hectares annually through reclamation would involve annual costs of close to $1 billion. A programme for recovering land lost over the last 25 years would cost in the region of $20 to $25 billion, spread over a period of 25 years, with recovery of the lost land achieved in 40 to 50 years.[10]

III   THE NEED FOR ADDITIONALITY AND AUTOMATICITY

The dynamics of world population growth and food requirements place on the international agenda for the next two decades one topic of immense urgency: the need, by the year 2000, to double the world's food-producing capacity. This implies that the food-producing land of the world will have to be expanded by some 70 million hectares annually, that the equivalent of this expansion must be achieved by increasing the productivity of existing agricultural land, or that both efforts should be undertaken in combination. The technical and financial resources required for doubling the world's food-producing land area in the next 20 years are massive. It is imperative that there be intensified efforts to halt the further deterioration of present land productivity, to stop the loss of food-producing land, and to reclaim areas lost through desertification during recent years. This is the main objective of the Plan of Action to Combat Desertification. The capital resources required are of the order of $1 billion per year.

In areas affected by desertification, measures aimed at halting further ecological degeneration, managing the resources of arid and semi-arid lands and reclaiming desertified land are integral parts of national development efforts. Existing and prospective flows of development assistance are not sufficient, however, to ensure either adequate rates of economic growth or the meeting of basic human needs in respect of health, shelter, nutrition and education. Consequently, it is not realistic to expect that either general developmental goals or the specific objectives of the Plan of Action will be achieved with existing or prospective resources. An initial stage of carrying out the plan can and should be made with these resources, but implementing it on any significant scale will require substantial infusions of additional funds.

Successful execution of the Plan of Action will also require a reliable flow of funds over a considerable period of time. To the extent possible, this flow should not be subject to the changing policies of individual governments, but should be automatic. Automaticity is important because an anti-desertification programme calls for a sustained, long-term effort, and such an effort must be based upon assured and predictable sources of finance.

The need for automatic sources of finance for international purposes is not a new concept. The Committee for Development Planning, at its Sixth Session in 1970, had a proposal for a tax on a limited number of consumer durable goods, the possession of which was indicative of the attainment of a high level of living by the purchaser. The contribution would be collected by the tax authorities of each country and used for financing international development.[11] The United Nations General Assembly declared during its Seventh Special Session in 1975 that the flow of concessional financial resources to the developing countries should be made 'predictable, continuous and increasingly assured'.[12]

More recently, the representative of Saudi Arabia proposed to the General Assembly during its Thirty-First and Thirty-Second Sessions the introduction of international taxation to provide additional funds to meet global environmental and other needs. He specifically suggested a tax of 1 cent per barrel on oil and proposed a voluntary tax of 0.1 to 0.2 per cent on arms sales.[13]

IV SOURCES OF FINANCING

Measures to combat the spread of deserts and to reclaim desertified lands will ordinarily be carried out by national governments as part of national programmes for development, and by groups of governments as regional co-operative programmes. These programmes will usually require external as well as domestic funding, and so it is necessary to identify the various sources of external funds which can be tapped, not only for global projects, but also for national and regional programmes.

International ventures, especially global ones of a programme of research or field experiments, will also have to be devised with the object of benefiting directly the affected countries. Donor countries which have contributed bilaterally in the past to specific international projects, as well as to national and regional programmes, will, it is hoped, continue their assistance. In paragraph 9 of resolution 32/172, the General Assembly called on developed countries, multilateral financial institutions and non-governmental donors to provide and increase their assistance to countries suffering from desertification, especially for the financing of their subregional and regional programmes and projects. The group considers that funding should also be sought from the private sector in developed countries and certain developing countries, and from the regional development banks.

The section which follows will deal exclusively with external (as opposed to domestic) sources of financing for the Plan of Action, including national, regional and international programmes and projects. It will deal first with funds supplied or raised specifically for anti-desertification projects and programmes, and then with the question of a share of new sources of financing established for general development and environmental purposes which may be allocated for anti-desertification purposes. The study presents below possible sources of financing.

A Funds provided expressly for anti-desertification programmes

Assistance from countries affected by desertification

Some of the countries affected by the process of desertification, both developed and developing, have ample financial resources and are exporters of capital. These countries should be invited to contribute funds for anti-desertification programmes in other countries on a grant or loan basis. In view of their experience and knowledge, as well as their capital resources, these countries may, in fact, be expected to take the initiative in accelerating the global effort to combat desertification.

Official development assistance

Official development assistance from members of the Development Assistance Committee of the Organization for Economic Co-operation and Development (OECD) amounted in 1976 to $13.7 billion, i.e., only 0.33 per cent of their total Gross National Product (GNP). This represented a decrease from the 1975 volume of $13.6 billion, or 0.35 per cent of GNP, and was still far short of the 0.7 per cent agreed to by most Committee members, and adopted by the General Assembly as part of the International Development Strategy for the Second United Nations Development Decade (resolution 2626 (XXV) of 24 October 1970). As a percentage of GNP, official development assistance has decreased substantially from the 1965–1967 average of 0.42 per cent.[14] It should be noted that while certain members of the Development Assistance Committee have reached or even exceeded the 0.7 per cent target, others continue to fall far short of it. Every effort should be made to increase official development assistance, but this does not appear to be an adequate and reliable source of financing to combat desertification. Furthermore, such assistance lacks automaticity, since it is voluntary or subject to each donor country's constitutional processes.

Nevertheless, recent developments in some large donor countries suggest a change in attitude which should be reflected in increased levels of official development assistance. The General Assembly, in resolution 32/181 of 19 December 1977 on the acceleration of the transfer of real resources to developing countries, has again called for an increase in the levels of such assistance. Given the importance of anti-desertification activities, it is to be expected that donor countries will agree to assign a high priority to this area in the establishment of their expenditure guidelines. Requests from countries will need to demonstrate this priority. As official development assistance increases, the funds specifically earmarked for combating desertification should be clearly identified by assignment of a separate expenditure category in this area.

Loans from national Governments and world capital markets

Because the governments of most developing countries lack the necessary funds, government lending must come from the developed countries and from developing countries with substantial financial assets derived from petroleum and other natural resources now in great demand. These loans would have to be granted on concessional terms, since:

(a) anti-desertification projects often have long gestation periods and benefits are not necessarily in the form of cash flows applicable to the repayment of principal and interest;
(b) most developing countries affected by desertification have limited debt-servicing capacities.

Established international financial institutions mobilize resources by selling securities in world capital markets, and they could raise additional resources in this manner to help finance anti-desertification programmes. The funds thus obtained, however, would involve commercial terms in their interest rates and repayment periods, and would not, therefore, meet the development financing requirements of the majority of the countries, which will need outright grants. This method of financing, therefore, would be appropriate for a few relatively advanced developing countries. Since anti-desertification projects often require funds on a concessional basis, it would be necessary even in such special situations to blend the funds obtained in the capital markets with grant funds and long-term low-interest development funds, in order to obtain the conditions required.

In a few special situations, anti-desertification projects will produce cash incomes or increases in land values that could be taxed to yield funds which would better enable the governments concerned to service external loans. As has already been noted, however, the benefits of anti-desertification measures often do not appear in the form of readily identifiable and taxable cash flows. Moreover, in many affected countries, especially in the Sahel, rangeland valuation is contrary to established customs. In any event, most of the countries severely affected by desertification already face serious balance-of-payments difficulties. Their ability to service foreign loans would not be significantly increased by collecting betterment charges from beneficiaries of anti-desertification projects.

Equity investment

Private investment accounts for the largest international flow of funds, have played and will continue to play a useful role in the financing of development in developing countries. This type of financing, however, has limited purposes and is most forthcoming for projects with immediate and high rates of return for investors. It cannot, therefore, be looked upon as a substantial source of funds for anti-desertification programmes *per se*, which are mainly in the public sector. There are multipurpose programmes with anti-desertification aspects in which private sector participation could be useful. Similarly, there will be anti-desertification programmes with components in which the private sector can clearly play a useful role. Opportunities exist, for instance, in the marketing of livestock and agricultural products and in the use of advanced technology. An analysis could be made by country or sub-region of projects suitable for private or possibly private/public financing in the light of their prospective yield to the investor. A possibility exists for the mobilization of equity funds through the creation of a public international corporation (see Part III, section V of this volume) which would attract investments from countries with surpluses in their international accounts, as well as from donor institutions, and would provide long-term financing for suitable projects with non-commercial rates of return.

Foundations

Foundations have also played a useful role in assisting development. Their resources, however, are limited and are usually devoted to special purposes, mainly in research and training. Foundations have been successful as catalysts in these areas, and should be encouraged to participate in financing training and research programmes, for example in the development of drought-resistant crops, solar stoves, sand-dune fixation and the like.

B Share of new sources of financing established for general development purposes[15]

The problem of financing international needs is not new. It has long concerned the international community. Two broad kinds of needs are visible:

(a) the need to transfer resources to the developing nations to assist them in improving the living standards of their people;
(b) the need to meet the financial requirements of various agreed international programmes.

The practice of depending on voluntary contributions to meet both kinds of needs has already proven to be inadequate and to carry with it a strong element of uncertainty. Resources should be transferred on a basis that is automatic and predictable.

As world attention is focused in turn on new global problems which only a world-wide effort can solve, the volume of funds required to finance internationally agreed programmes grows steadily. The decision to embark upon a global programme of anti-desertification measures provides further reasons not to delay the working-out of a system for the automatic generation of resources for general financing purposes. The fact that the level of assistance from developed countries has not met the voluntary target of 0.7 per cent of gross national product points to the need for more systematic mechanisms entailing an element of automaticity. Such an element would help ensure a more dependable source of revenues from which developing countries, including those suffering from desertification, can finance their development programmes.

Revenues from exploitation of the international commons

(a) *Non-living ocean resources*
The General Assembly has declared the non-living ocean resources beyond national jurisdiction to be the 'common heritage of mankind'. [See *United Nations Convention on the Law of the Sea*, 1982.] The Third United Nations Conference on the Law of the Sea has, however, thus far been unable to agree on the nature and powers of an international regime to control the exploitation of these resources. Moreover, the extent of the resources available for international exploitation is being reduced by the assertion by coastal States of exclusive economic rights over zones extending 200 miles or more from land.

The only promising known resources that lie beyond the Exclusive Economic Zones (EEZ) claimed by coastal States are the manganese nodules found on various parts of the ocean floor. A number of large firms (principally American, but also Japanese and European) have shown an interest in mining these nodules for their nickel, copper and cobalt content, and, in the case of some firms, also for their manganese content. Actual mining operations on a commercial scale have not begun, so their profitability remains unknown. It is possible, however, that a nodule mining industry might produce significant economic rents (i.e. net revenues after costs and a normal return on capital) of several hundreds of millions of United States dollars annually by the mid or late 1980s.

The economic rents might in principle be appropriate for international purposes. Not all of them, however, would necessarily be available for development or environmental purposes. Care of the ocean may seem to have a prior claim on such funds, and some of them may be used to finance the proposed international sea-bed authority. Others might be used to compensate countries whose land-based mining industries would be damaged by the new nodule mining industry. Although other activities might have a prior claim on these resources, the global anti-desertification effort could also claim a share.

The resources of the claimed exclusive economic zones appear to be much richer than those of the deep ocean. Large revenues will probably be produced through exploitation of oil and gas deposits within those coastal zones. The Third United Nations Conference on the Law of the Sea should not legitimize those zones without at the same time asserting a claim on behalf of the international community to some percentage of the rents that those zones will yield. The percentage should vary with the levels of development of the coastal States. Here also, anti-desertification programmes could claim a share of this international revenue.

(b) *Taxes on polluters of the marine environment*
The most pervasive pollutants of the marine environment are crude oil and refined petroleum products. In principle, a tax on the amount of oil or refined products discharged into the ocean would induce polluters to reduce their polluting emissions and, to the extent that total elimination of emissions was uneconomic, produce some revenue for international purposes. In practice, a tax that varied closely with the volume of pollutants emitted could not easily be administered. A more practical alternative would be to tax polluters who installed anti-pollution technology at a lower rate than those who did not.

For example, tankers with segregated ballast tanks do not on the average discharge as much oil into the ocean as those that are not so equipped. A differential tax on these two categories of tankers could induce owners to order new tankers with segregated ballast tanks when old tankers are retired. In the first year, such a tax might yield about $30 million. In the twentieth year, assuming that all tankers would by then have segregated ballast tanks, revenues would fall to about half that amount. Average annual revenue over the 20-year period would be something more than $20 million. As in the case of revenues from non-living ocean resources, a good part of the funds generated could be set aside for the care of the oceans, but again anti-desertification programmes could have a share of funds generated through such measures.

(c) *Telecommunications and satellites*
Two special attributes of the physical universe can properly be regarded as part of the international commons: telecommunications wavelengths, and parking spaces or 'slots' for satellites in geosynchronous orbit. Economic rents produced by exploitation of these physical phenomena could, in principle, be appropriate for international purposes.

The use of telecommunications channels presumably already produces significant rents, but appropriating them for international purposes would encounter serious complications. The assignment of frequencies is largely in the hands of national governments, and the profits of the firms using those frequencies are already subject to national taxation. Moreover, an increasing part of the long-distance use of telecommunications frequencies involves satellites belonging to the International Telecommunications Satellite Organisation (INTELSAT), which already charges for the use of its circuits. A less difficult variant could be a small national surtax on all telephone and telegram/telex charges, the proceeds of which could be used for international purposes.

Charging for parking spaces or 'slots' for satellites in geosynchronous orbit may one day become a source of revenue for international purposes. These slots are not yet a scarce resource, but the time may come when international action will be needed to control the allocation of the choice slots that are all well located with respect to economically-important land areas. A good case can then be made for charging for the use of such slots.

International taxation of trade flows

In 1978, total world trade should reach $1,000 billion. A tax of only 0.1 per cent on the value of international trade, therefore, would yield $1 billion annually. The primary attraction of a general trade tax is the volume of revenue it would generate, and it can be kept at a low rate. Consequently, it would be less burdensome than other sources of revenue, and need not be a disincentive to world trade. Trade can be viewed as an appropriate object of international taxation, since it depends on world order and often on the use of the international commons.

The tax would be paid directly by firms and individuals, including public trading corporations, after Governments had given their initial assent. This would give it a certain automaticity. The tax could be levied at either the point of importation or the point of exportation. Various adjustment factors to mitigate likely difficulties will need to be explored.

Special drawing rights link

Various proposals have been considered by the international community from time to time to establish a link between the use of special drawing rights (SDR) as the principal asset in the international reserve system and the provision of finance to developing countries. It is agreed that the total volume of SDR allocated should be determined exclusively by global liquidity needs. The decision to establish the link has been delayed up to now by a conflict of interests and perceptions. Those developed countries favourable to the link wish to ensure that the resources transferred will benefit primarily the least developed countries. Many developing countries wish to ensure that all benefit from the link. All the developing countries would prefer a different system for allocating new issues of SDR from the present one, which is based on the global distribution of International Monetary Fund (IMF) quotas.

Under the present conditions of the world economy, a new issue of SDR would stimulate demand in a large number of industrialized countries, give a boost to world trade and move the system away from excessive reliance on any single reserve currency.

IMF should work out a formula for the allocation of new SDR which benefits all countries. The largest benefits would go to the least developed countries, with gradually decreasing benefits for the middle- and high-income countries.

Proposals to issue a portion of the new supply of SDR to specific international institutions have been made in the past. It is now proposed that whichever institution is given the responsibility to manage global anti-desertification funds be allocated a substantial portion of any new issue of SDR to help finance measures to combat desertification.

Military expenditures

(a) *Savings from disarmament*
Since the inception of the League of Nations, disarmament has been a continuing subject of international debate and aspiration. For more than 20 years the United Nations has discussed disarmament with a view

to promoting international peace and security and providing funds for the development of the poorer countries of the world. The General Assembly, in resolution 3093 A (XXVIII) of 7 December 1973, recommended, *inter alia*, that all States which are permanent members of the Security Council reduce their military budgets by 10 per cent from the 1973 level during the next financial year, and appealed to those States to allot 10 per cent of the funds released for assistance to developing countries. The present level of world military expenditure is currently estimated at $360 billion annually, a level of use of resources several hundred times the estimated volume of funds needed to halt the spread of deserts. A reduction of merely 5 per cent could release some $18 billion a year. If one half of this amount could be earmarked for international programmes of development and environmental protection, including an effort to combat desertification, over $9 billion a year would be available for these purposes. An intensified effort to reverse the present trend in arms expenditure is imperative.

(b) *Taxation of military expenditures*
This tax would be imposed on all military expenditures. The basis would be either the provisions made in the national budget, or sums actually disbursed, as reported by the government to a central authority. So as to avoid some of the reporting difficulties which are likely to be encountered in efforts to tax current military expenditures, the expedient might be adopted of basing the tax on reported expenditures for an earlier period: for example, expenditures incurred three years previously. Countries would probably have less difficulty in reporting accurately what they spent three years before. A tax of one tenth of 1 per cent on expenditures reported would be too small to have much effect on the total military burden, but it would yield an annual income of the order of $300 to 350 million.

(c) *Taxation of arms transfers*
Since the general purpose of this tax would be to discourage the traffic in arms, it could be imposed at a rate substantially higher than the tax on military expenditure as a whole. Consideration should be given to imposing a tax of 5 per cent on all arms transfers. Again there should be no exemptions.

Since 1973, the world total of arms transfers has been running at a level between $9 billion and $10 billion annually. A tax at the rate of 5 per cent on all such transfers should yield between $450 million and 500 million. If it is assumed that a tax at this rate would have the effect of reducing the volume of sales by, say, one tenth, the annual yield would still be of the order of $400 million. The tax in this case would be collected from the supplier country.

Others

Among other possibilities, the group has noted that in the context of the negotiations regarding a Common Fund at present taking place under the auspices of the United Nations Conference on Trade and Development (UNCTAD), proposals have been advanced which envisage the possibility of measures being taken through the Fund for purposes other than the buffer stocking of commodities. It is hoped that, should any agreement be reached on these proposals, account will be taken of the financial needs related to the implementation of anti-desertification programmes by developing countries suffering from desertification.

The problem of burden-sharing

The various possible sources of finance discussed above will not impose equal burdens on all countries, either in total magnitude or in relation to their gross national products. This fact is not in itself a reason for rejecting any of these sources of finance, since differential burdens are an unavoidable attribute of all workable revenue systems, both national and international. Nevertheless, some adjustments of national burdens would clearly be required in the interests of equity and political acceptability. Thus, the burdens imposed on individual countries should, in principle, be directly related to their level of development. This result could be achieved by permitting poor countries to retain part or all of the funds that they collect for international purposes (e.g. in the case of trade taxes), or by selecting sources of revenue which, by their nature fall more heavily on rich than on poor countries (e.g. the SDR-link). The distribution of benefits from the disbursement of international revenues should also be taken into account in evaluating the fairness of the measures used to create those revenues.

V THE MOBILIZATION AND MANAGEMENT OF FUNDS

The mobilization and management of funds are essential elements of the measures for financing the Plan of Action to Combat Desertification. In the case of funds raised or contributed directly for anti-desertification programmes, responsibility for their mobilization and co-ordination should remain with the Governing Council of UNEP, as the body entrusted with supervision of the implementation of the Plan. The consultative group which the General Assembly has authorized the Executive Director to convene (resolution 32/172, para. 10) can play a particularly useful part, through its periodic meetings of donors and recipients, in raising levels of assistance. If the public international corporation[16] envisaged under Equity Investment (page 11) is indeed established, UNEP would have this new institution available to assist in the raising and management of funds obtained from the capital markets and through equity investments. It is assumed that funds contributed specifically for anti-desertification programmes and projects (except *ad hoc* funds-in-trust) will be deposited in a special account and managed on such lines as may be approved by the General Assembly.

The establishment of new sources of financing for general international purposes, of which the programme to combat desertification would have a share, is a more complicated problem. A number of consequences would arise from a decision by the international community to establish new general sources of funds to finance global programmes of development or environmental protection, involving the principle of automaticity:

(a) The first is that the provisions for the generation and collection of the funds will have to be agreed upon by international treaty.
(b) Secondly, the governments will, except in the case of the SDR link, have to act as agents for the collection of funds and for channelling them to some central international authority.
(c) Thirdly, there will be need for a central international mechanism for the collection of funds centrally.
(d) Fourthly, some kind of policy board would be needed for allocating the funds among the different agencies, sectors and fields of activity.
(e) Finally, an appropriate intergovernmental body will have to be assigned the responsibility for the programming and expenditure of the share of such general development funds allocated for anti-desertification measures.

In some cases, the collection of the new sources of revenue discussed in this study could be assigned to existing mechanisms. Thus, national customs offices could collect international taxes on trade, and national port authorities could enforce pollution charges on shipping. In other cases, new mechanisms would be necessary. For example, the proposed international sea-bed authority would seem to be the proper agency for collecting revenues from the exploitation of non-living resources of the ocean beyond national jurisdiction.

The allocation of the new resources established for general development purposes should be entrusted to a broadly based and high-level intergovernmental policy body with specific responsibilities for the promotion and co-ordination of economic and social development, such as the Economic and Social Council. The funds may be allocated according to the broad international programmes to be financed, e.g.

for economic development, for social development and for environmental protection. The system of allocation might be determined additionally on the basis of the terms under which the funds are to be provided. In that case, funds could be allocated to the International Bank for Reconstruction and Development (IBRD) for programmes to be financed on commercial terms, to the International Development Association (IDA), the International Fund for Agricultural Development (IFAD) or a restructured Capital Development Fund for programmes to be financed on concessional terms, to the United Nations Development Programme (UNDP) for technical and pre-investment assistance, to the Fund of UNEP for environmental activities and anti-desertification programmes, and to other members of the United Nations family according to purpose and terms.

The deposit of funds in the special account would constitute, in effect, the establishment of an international fund to combat desertification, which was among the measures suggested by the United Nations Conference on Desertification for inclusion in the present study (para. 104 (e) of the Plan of Action). Such a fund (or special account) could be managed by the Executive Director of UNEP under the policy guidance of the Governing Council, in line with the General Assembly's decision to entrust them with the responsibility of following up and coordinating the implementation of the Plan of Action (resolution 32/172, para. 8).

Funds-in-trust

The United Nations Conference on Desertification suggested that the group preparing the study should also consider funds-in-trust (para. 104 (e) of the Plan of Action). By their very nature, such funds are established for a specific purpose by agreement between the donor and the administering body. The arrangements discussed for the mobilization and management of funds should not preclude the entering into agreements by donors and administering bodies for the establishment of trust funds for specific purposes relating to desertification.

VI CONCLUSIONS

It is clear from this study that desertification is one of the world's major problems, involving extensive loss of human life, abject poverty and the degradation of a vital, life-sustaining part of the environment. The annual production losses due to the effects of desertification amount to nearly $16 billion. Expenditures on the order of $400 million annually would be needed merely to achieve 'zero desert growth'. Because of the seriousness of this environmental hazard and impediment to development, it is important not only that the spread of the deserts be halted, but also that formerly productive but now desertified land be reclaimed. A programme to reclaim lands desertified over the last 25 years alone would require financing of the order of $20 to $25 billion.

The additionality of resources is imperative for financing this programme. Because of the urgency of the problem, however, implementation of the Plan of Action should begin immediately with existing resources, rather than await the availability of additional funds. At the same time, the international community should explore both the possibility of mobilizing additional funds through existing financial mechanisms and the possibility of establishing new sources.

In considering existing sources, countries – both developed and developing – having ample financial assets should be asked to increase funds made available for anti-desertification programmes. Additional official development assistance should be channelled to combat desertification, which should be recognized as a distinct priority concern. In special cases, loans from Governments and funds raised in capital markets should be devoted to combating desertification, and private investment and assistance from foundations should also be used for this purpose.

In view of the magnitude and long-term nature of the anti-desertification programme, however, resources will be needed which are not only additional, but also predictable and dependable. New sources entailing automaticity should also be established for general development purposes, and part of these resources should be allocated to the anti-desertification programme.

The study considered a variety of possible fund-raising measures entailing automaticity. These include revenues from the exploitation of the international commons, taxation of defence expenditures and arms transfers, a development link with SDR, and the taxation of international trade flows. The establishment of these new sources could generate considerable funds, of a magnitude far surpassing those generated through any present means of raising development capital, and would thus go a long way towards meeting the needs of development and environmental protection. While the establishment and utilization of these resources are undoubtedly complex matters, the practical difficulties can be resolved, in the same way that they have been in the case of revenue-raising mechanisms employed by national governments. The question is one of political will – whether the United Nations General Assembly is prepared to pursue what is clearly the avenue with the greatest potential for the mobilization of funds for development. Even if this is in the nature of a longer-term proposal, the group believes that action should be initiated now.

Finally, the international community should recognize desertification as a distinct and priority concern, calling for massive, sustained action. The problem is a serious challenge to the nations of the world, and the struggle against it should be a major claimant on the world's financial resources, both existing and new.

Notes

1 World Map of Desertification (at a scale of 1:25 000 000) FAO, UNESCO, WMO, 1977, with explanatory note (A/CONF.74/2).
2 *United Nations Conference on Desertification:* Round-up, Plan of Action and Resolutions, UN Publication, New York, 1978.
3 See 'Transnational project on management of livestock and rangelands to combat desertification in the Sudano-Sahelian regions (SOLAR)' (A/CONF.74/26).
4 See 'Transnational green belt in North Africa' (A/CONF.74/25).
5 See 'Sahel green belt transnational project' (A/CONF.74/29).
6 See 'Transnational project: The management of the major regional aquifers in north-east Africa and the Arabian Peninsula' (A/CONF.74/24).
7 See 'Transnational project to monitor desertification processes and related natural resources in arid and semi-arid areas of south-west Asia' (A/CONF.74/28).
8 See 'Transnational project to monitor desertification processes and related natural resources in arid and semi-arid areas of South America' (A/CONF.74/27).
9 The National Council for Research, Khartoum, Sudan, 1976.
10 These and the preceding figures are estimates of orders of magnitude, prepared by the secretariat of the United Nations Conference on Desertification (see document A/CONF.74/3/Add.2). For refined estimates see Part II.
11 *Official Records of the Economic and Social Council, Forty-Ninth Session, Supplement No. 7* (E/4776).
12 Resolution 3362 (S–VII) of 16 September 1975, sect. II, para. 1.
13 Documents A/C.2/31/SR.45 and A/C.2/32/SR.23. The need for automaticity was also dealt with in some detail in a United Nations forum by the Secretary-General of the United Nations Conference on Desertification when presenting the Draft Plan of Action to Combat Desertification to the Conference's Committee of the Whole (DESCONF/CRP.3).
14 OECD, 1977 Review: *Development Cooperation* (Paris, November 1977).
15 Most of the financial measures discussed in this section are analysed at greater length by Eleanor B. Steinberg and Joseph A. Yager with Gerard M. Brannon in *New Means of Financing International Needs* (Washington, DC, Brookings Institution, 1978).
16 Described in Part III, section V.

List of participants in group of high-level specialists in international financing who prepared the study contained in Part I
(Serving in personal capacities)

Rodrigo Botero (Chairman)
Member, Independent Commission on International Development Issues
(Former Minister of Finance, Government of Colombia)

Chiu Wen-Min
Principal Officer, Office of the Secretary-General, UNCTAD

Kenneth Dadzie
Ambassador of Ghana to Switzerland and Austria and Permanent Representative to the United Nations Office at Geneva
(United Nations Director-General designate for Development and International Economic Co-operation)

R. K. A. Gardiner
Commissioner for Economic Planning,
Ministry of Economic Planning,
Government of Ghana
(Former Executive Secretary, Economic Commission for Africa)

Abd-El Aziz Mohamed Hegazy
Former Prime Minister and Minister of Finance,
Government of Egypt

Paul-Marc Henry
President, Society for International Development
(Former Assistant Administrator, UNDP, and President, Development Centre, OECD)

Felipe Herrera
Chairman of the Board, United Nations Institute for Training and Research
(Former President, Inter-American Development Bank)

J. Kaufmann
Ambassador of the Netherlands to Japan and former Permanent Representative to the United Nations;
President of the Governing Council of UNDP

C. Kerdoudi
Vice-President in Charge of Operations and Projects Departments,
African Development Bank

Fawzi Mahresi
Representative for Europe,
Arab Bank for Economic Development in Africa

C. S. C. Mselle
Minister, Permanent Mission of the United Republic of Tanzania to the United Nations;
Chairman, Advisory Committee on Administrative and Budgetary Questions

Sir Edgerton Richardson
Ambassador on Special Missions and Special Adviser on Foreign Affairs to the Prime Minister,
Government of Jamaica

Marice Strong
Chairman, Board of Governors,
International Development Research Centre
(Former Executive Director, UNEP)

Christopher Thomas
Deputy Permanent Representative of Trinidad and Tobago to the United Nations;
Member, Advisory Committee on Administrative and Budgetary Questions

Mahbub Ul Haq
Director, Policy Planning and Programme Review Department, World Bank

Ambassador Maurice Williams
Chairman,
Development Assistance Committee, OECD

Joseph Yager
Senior Fellow,
The Brookings Institution

CONVENOR
Mostafa K. Tolba
Executive Director, UNEP

SECRETARIAT
Secretary: *Ruben P. Mendez*
                Principal Officer, UNEP

Assistant Secretary: *Nadjib-Ullah Omer*
                       Programme Officer, UNEP

UNEP SECRETARIAT
Noel Brown
Director, UNEP Liaison Office,
New York

CONSULTANT
M. A. Kassas
Professor of Ecology,
Faculty of Science,
Cairo University

PART II

Study on Financing the United Nations Plan of Action to Combat Desertification (1980)

*Original Text:
General Assembly document
A/35/396 of 17 September 1980*

I INTRODUCTION

The study was prepared in accordance with General Assembly resolution 34/184 of 18 December 1979. In that resolution, the General Assembly, *inter alia*, expressed concern over the lack of adequate financial resources for the implementation of the Plan of Action to Combat Desertification (PACD) and, as part of its search for a solution to this problem, requested the Secretary General, in consultation with the Governing Council of the United Nations Environment Programme (UNEP), to submit to the General Assembly at its Thirty-Fifth Session a report on this subject. The General Assembly specified that the report should be based on a study prepared by a group of high-level specialists in international financing to be convened by the UNEP Executive Director and that the study should deal with the following:

(a) A complete inventory of relevant ideas and proposals put forward in the United Nations System of possible new ways and means to finance programmes of multilateral organizations at the world level, additional to regular assessed budgets and conventional extra-budgetary resources;
(b) A financial plan and analysis outlining the components and costs of a programme to stop further desertification and identifying what is already being financed and what additional resources may be needed to meet the minimum objectives of stopping the spread of desertification;
(c) Methods for the mobilization of domestic resources;
(d) The practicality of obtaining loans from Governments and world capital markets on a concessionary basis;
(e) The feasibility of the creation of a public international corporation which would attract investments from countries as well as institutions and would provide financing for suitable anti-desertification projects with non-commercial rates or returns;
(f) The means for encouraging the active participation of foundations in the financing of anti-desertification training and research programmes.

The group of high-level specialists in international financing met in Geneva from 21 to 25 July 1980. The group considered and adopted a draft of the study prepared by a core group of its members, assisted by a substantive secretariat.

The order of chapters in the study is in accordance with the sequence of subjects as listed in paragraph 7 of resolution 34/184. In chapter I, an attempt is made to present a complete inventory of all proposals and ideas presented in the United Nations system for financing programmes of multilateral organizations additional to regular assessed budgets and conventional extrabudgetary resources. It should be noted that the inventory has been prepared according to a functional classification, rather than in order of importance. Various references have also been made to the report of the Independent Commission on International Development Issues under the chairmanship of Willy Brandt. The implications of their report, *North-South: A Programme for Survival*, 1980 (hereafter referred to as the Brandt Commission Report), are considered to be of such significance as to warrant the inclusion of the relevant recommendations in the inventory.

Summary of findings

Three years ago [1977], the United Nations Conference on Desertification (UNCOD) adopted a plan of action to combat the loss of productive land to spreading deserts. Only limited, and quite inadequate, resources have been mobilized to carry out the PACD. Desertification, with its serious social and economic consequences, continues to be a major world problem.

About one-third of the earth's land surface (excluding Antarctica) is arid or semi-arid. Eighty per cent of the agricultural land in arid and semi-arid regions is affected by desertification which progressively decreases the land's productivity. Every year over 20 million hectares of land deteriorate to the level at which they yield zero or negative net economic returns.

Over half of the land affected by desertification is in countries that appear able to finance their own anti-desertification programmes and pay for the technical assistance they may require. The rest of the affected land is in less favourably situated countries, including some of the poorest countries in the world. A programme designed to stop the further loss of productivity caused by desertification in these countries would require average annual expenditures on the order of $1.5 to $4 billion, with a medium estimate of $2.4 billion over a period of 20 years. These expenditures would be devoted solely to the restoration of the productivity of affected land. Additional expenditures, which cannot now be estimated, would be required for other essential measures, such as providing alternative livelihood systems and sources of energy to the populations involved.

The financial plan to stop desertification presented in this study calls for restoring the productivity of 100 per cent of the affected irrigated land, 70 per cent of the rainfed cropland and 50 per cent of the rangeland. Gains in productivity on irrigated land would compensate for the cost, but restoration of productivity of all affected rangeland and rainfed cropland could probably not be justified on economic grounds.

Most of the funds required to carry out the financial plan would have to be raised externally. How much external assistance is available for combating desertification, as opposed to relief and general development, cannot easily be determined, but it is clearly not enough. Total external assistance involving desertification control has been estimated at about $500 million in 1978.

A substantial part of the resources required in anti-desertification projects is domestic. It does not follow, however, that requirements for external assistance are correspondingly reduced. Income levels in many countries severely affected by desertification are very low. The fiscal capabilities of their governments are weak, and institutions for accumulating and investing savings are limited or non-existent. Expenditures by developing countries in 1978 to combat desertification were on the order of only about 10 per cent of total expenditures required for that purpose, including external assistance.

Some domestic resources might be mobilized by increasing tax collections or diverting revenues from lower priority uses. And as the anti-desertification programme begins to produce results, charges can be levied on the use of land whose productivity has been increased. At best, however, the affected developing countries could be expected to raise only a small fraction of the funds needed to stop desertification.

In many countries, a large proportion of the domestic resources required for anti-desertification work – principally labour – can be mobilized only by an injection of foreign assistance. One approach would be to provide goods to be sold on the local market and put the resulting proceeds in a special account to finance anti-desertification projects. Another method would be to ship in food to pay for labour on anti-desertification projects, as in done by the World Food Programme. Still another possibility is to organize joint enterprises to raise cash crops, including new varieties, suitable to areas with limited rainfall. It would be hoped that the foreign capital and technical assistance put into such enterprises would stimulate the investment of domestic capital.

As previously stated, the medium estimate of the average annual cost of the programme in support of the anti-desertification effort is in excess of $2.4 billion, of which it has been estimated that about $500 million is currently (1978) being provided from external sources. This leaves a requirement from external sources of about $1.8 billion annually. The programme would be smaller in the early years of the effort than it would be later. Some of the most urgent requirements of the affected areas would, however, have to be met in the early years.

At first, requirements for external assistance will of necessity be met exclusively from conventional sources, principally grants and concessional loans from governments and international financial institutions. Experience with other assistance programmes indicates that obtaining grants and concessional loan funds will not be easy. As the scale of the anti-desertification effort grows other sources of finance will have to be developed.

Some anti-desertification projects (e.g. rehabilitation of irrigated farmlands) yield higher rates of return that would appear to make them eligible for loans from international capital markets. However, the governments of the poorer countries affected by desertification have limited experience in international finance, and their ability to service foreign loans is limited, especially if the terms are not concessional. The services of an intermediary to gain access to capital markets and to arrange concessional terms would be necessary. The intermediary could be an existing institution, such as the World Bank or one of the regional development banks. If existing institutions are unwilling, or unable by their statutes, to increase their activities in the field of desertification control, a public international corporation might be created specially to arrange financing for anti-desertification projects. The principal purpose of such a corporation would be to provide funds on highly concessional terms.

A public international corporation could be chartered independently by a group of interested governments, or it could be established as an affiliate of an existing institution. In the former case, the corporation could contract with one or more existing institutions for technical services on a reimbursable basis. If the corporation were an affiliate, it could be put under the board of directors of the parent institution, or it could be governed by its own board of directors.

The ability of the corporation to borrow in capital markets could be assured if a number of industrialized and oil-exporting governments provided equity capital and guaranteed its obligations. The corporation could achieve concessional terms in its anti-desertification lending operations by blending money borrowed on commercial terms with grants and concessional loans from aid donors or with funds raised by one of the new means described in this study. Alternatively, it could use grants, concessional loans, or new revenues to cover both its administrative costs and part of the interest charges on money borrowed in capital markets. Some of the corporation's equity capital could also be used for blending and to hold down interest rates charged on anti-desertification loans.

Borrowing in capital markets could be only a small part of the answer to the problem of financing the PACD. The ability of the affected developing countries to service even concessional loans is limited. Moreover, the total volume of loans would be constrained by the guarantees that sponsoring governments would be prepared to extend to cover borrowings in capital markets.

Grants from private foundations for research and training programmes have been suggested as an additional source of funds. Training and applied research are important parts of the effort to control desertification. Foundations are not now active in these fields, but it would be desirable to encourage their participation in the future. Governments that do not now do so should be urged to provide tax incentives to private donors that could stimulate foundations to play a greater role in international training and research programmes. At best, however, foundations could be expected to provide only a small part of total financial requirements.

If an adequate anti-desertification programme is to be carried out, new means of finance will have to be found. When account is taken of future needs in the field of general development, the necessity of finding new means of finance is all the more compelling.

New means of finance should be administratively feasible, create no severe inequities, and possess a degree of automaticity. Among the many new financial measures that have been proposed in the United Nations system, several seem to meet these criteria. These measures are of three kinds: international financial measures, revenues from the exploitation of the international commons, and taxes on international trade. All of these measures are potential means of financing any recognized international need. Desertification control would be one important claimant to the funds that these measures could generate.

In the field of international finance, three measures deserve serious consideration: the resumption of gold sales by the International Monetary Fund (IMF), the creation of a link between the special drawing rights (SDR) issued by the IMF and assistance to developing countries, and the establishment of a Common Fund to stablize commodity prices.

(a) Between 1 July 1976 and 30 June 1980, the IMF sold 25 million ounces of gold and returned a similar amount of gold to its members. The profits of gold sales were put in a trust fund from which concessional loans were made to poor countries experiencing balance-of-payments problems. The IMF still holds approximately 100 million ounces of gold. If gold sales are resumed, the need to combat desertification should be made a factor to be considered in making loans from profits. If additional gold is returned to members, they should be encouraged to use part of it to finance anti-desertification projects.

(b) SDR are issued only when the international monetary system requires additional liquidity, so they are not a reliable source of finance for anti-desertification projects or other international needs. Moreover, they are distributed in proportion to the quotas of IMF members, which results in their going largely to wealthy countries. SDR could be made available to help poor countries for

anti-desertification and general development in one or two ways. They could be distributed in accordance with a new allocation formula favouring those countries. Or they could be issued in whole or in part to an international institution that would make loans to poor countries for anti-desertification and other purposes.
(c) The recent Agreement to establish a Common Fund to stabilize the prices of commodities subject to international commodity agreements could also benefit countries suffering from desertification. Those countries could apply for assistance from the Common Fund's 'Second Window'.

The international commons consist of the oceans beyond national jurisdiction and outer space.

(a) On the basis of discussions at the Third United Nations Conference on the Law of the Sea, the oceanic part of the commons can be divided into three zones: the area lying between 12 and 200 miles from shore, the continental shelf where it extends more than 200 miles from shore, and the deep ocean. The first two areas will probably be designated exclusively economic zones for exploitation by coastal States. Before legitimizing the claim of the coastal States to the resources of these areas, it would be proper for the Conference to require that a fraction of revenues derived be made available for international purposes. At present (1980), the Conference appears to be considering applying this requirement only to those parts of the continental shelf more than 200 miles from shore. Some of the revenues derived from exploiting the non-living resources of the deep ocean – principally manganese nodules – will probably be made available for international purposes, but it is difficult at this point to estimate how large those revenues will be.
(b) The only use of space that might be a source of revenue for international purposes in the near future is the 'parking' of geostationary communication satellites in a narrow band circling the earth at the equator. This band is becoming crowded, and parking spaces may have to be allocated. If so, charges might appropriately be imposed for their use.

A general trade tax could be a powerful revenue-raising measure. A 0.5 per cent levy on the value of all international trade would yield about $7 billion annually. Even after exemptions for very poor countries and adjustments for countries especially dependent on international trade, the revenue would be quite large. Smaller, but still substantial, revenues could be raised by imposing low taxes on trade in specific commodities, such as energy materials and exhaustible minerals.

The co-ordination of the various sources of funds to finance the

PACD is a complicated problem. Some funds will flow directly from aid donors or international financial institutions to anti-desertification projects. Other funds will be channelled through the *Special Account* administered by UNEP to finance the Plan of Action to Combat Desertification, the United Nations Trust Fund for Sudano-Sahelian activities, or the public international corporation, if it is established. Any new, unconventional sources of finance would have to be allocated among various claimants by some existing institutions, such as the World Bank or the UNDP, or by a new institution created for the purpose. The funds from such new sources allocated to anti-desertification activities would then flow to projects directly or by way of the *Special Account*, the United Nations Sudano-Sahelian Office (UNSO) trust fund, or the public international corporation. Co-ordination and, if possible, simplication of these various financial arrangements would obviously be desirable.

II INVENTORY OF MEANS OF FINANCING INVOLVING AUTOMATICITY PROPOSED IN THE UNITED NATIONS SYSTEM

Background

While the proposed inventory of new ways and means to finance multilateral programmes other than 'regular assessed budgets and conventional extrabudgetary resources' has been requested by the General Assembly as part of a special study on the financing of the PACD, this section clearly has implications which go far beyond the subject under consideration in the Plan of Action. These implications touch on the fundamental question of the nature of development assistance itself, particularly its adequacy, reliability and predictability, all of which are affected by monetary factors.

In various fora, both within and outside the United Nations system, serious doubts have been expressed regarding the adequacy of conventional extra-budgetary resources for financing international as well as bilateral programmes. These resources take the form mainly of voluntary contributions. Although they have been a significant and effective source for financing development, it has been noted that this type of assistance is often tied and, being of a voluntary nature, is subject to the vagaries of national political conditions. As is well known, the generally accepted target of devoting 0.7 per cent of GNP to official development assistance is far from being reached, the actual percentage global amounting to 0.34 per cent in 1979 [now (1986) even less]

The question of assessed budgets as a source of financing multilateral assistance programmes has not been the subject of as extensive discussion as has development assistance, primarily because they are basically not intended for multilateral assistance but rather for the administrative and operating costs of international organizations. With the exception of the World Health Organization (WHO), only small percentages of assessed budgets are used for assistance programmes *per se*. It has also been noted, however, that even under a system of assessed budgets, the independence, operation and financial viability of the organizations of the United Nations system can also be subject to the vagaries of national political conditions, which is less likely to be the case if these organizations had independent sources of income.

In its request for 'a complete inventory of relevant ideas and proposals of ways and means to finance programmes of multilateral organizations at the world level, additional to regular assessed budgets and conventional extrabudgetary resources', the General Assembly thus had in mind the need for a new approach to financing. During its Seventh Special Session, the General Assembly declared, in resolution 3362 (S-VII) of 16 September 1975, that the flow of concessional financial resources to the developing countries should be made 'predictable, continous and increasingly assured'.

The PACD, which was adopted by the United Nations Conference on Desertification (UNCOD) held at Nairobi from 29 August to 7 September 1977, had recommended a study of additional measures of financing, including 'fiscal measures entailing automaticity' and the General Assembly, in approving the Plan of Action in its resolution 32/172 of 19 December 1977, specifically invited the UNEP Governing Council to have such a study prepared. (See Part I.)

Although numerous ideas and proposals have been made on this subject, the General Assembly has requested an inventory of ideas and proposals put forward in the United Nations system. The inventory will therefore cover, except for references to the Brandt Commission report (see introduction, page 27), only those proposals and ideas which have been presented in the United Nations, including its principal organs, subsidiary organs and specialized agencies as well as United Nations conferences, expert groups, *ad hoc* committees, and other fora held under the auspices of the United Nations system.

The subject of generating resources for financing international purposes on an assured and predictable basis is not new. In fact, it has surfaced recurrently in the United Nations system in various forms for some time now. For instance, the idea of compensatory financing or buffer stocking to stablize commodity prices – involving automatic compensatory schemes – was examined within the United Nations system in the early 1950s in a report, 'Commodity Trade and Economic

Development' (E/2519), prepared by a Committee appointed by the Secretary-General at the request of General Assembly resolution 623(VII) of 21 December 1952, on the subject of fluctuations in the prices of primary commodities and their relation to the prices of capital goods and manufactured products. Similarly, the idea of the use of sea-bed resources 'in the interests of mankind' was discussed extensively as early as 1967 by the General Assembly, which established an *ad hoc* Sea Beds Committee to pursue the matter.

It was not, however, until UNCOD (1977) that the principle of 'automaticity' as such in international financing was explicitly presented in a United Nations forum. The Secretary-General of the Conference, in addressing the Conference's Committee of the Whole noted that the implementation of the PACD was a long-term process which required international assistance on a sustained, predictable and assured basis. In the deliberations that followed, this issue was extensively discussed, and the Plan of Action adopted by the Conference specifically requested that a study be undertaken of additional measures and means of financing for the implementation of the PACD including 'fiscal measures entailing automaticity'. The concept of 'automaticity' was put forward to distinguish it from the present general system of financing on a purely voluntary basis.

In view of the wide variety of proposals which have been made for automatic sources of financing, a classifiation of the proposals has been prepared, based on broad categories of fiscal measures involving automaticity. These broad classications are: (a) international taxation; (b) income from the use of the international commons; (c) income from military taxes and savings from disarmament; (d) monetary measures such as an SDR-development link, and a link between gold sales and development; (e) commodity prices stabilization mechanisms and orderly commodities development; and (f) other miscellaneous proposals.

It should be noted that as this is an inventory it includes all proposals and ideas regardless of their stage of advancement. Many of the proposals have been presented informally, others in the form of draft resolutions, or resolutions adopted by the United Nations General Assembly, United Nations conferences, or governing bodies of other United Nations entities.

In a few instances, the proposals are close to realization or have already been initiated. The inventory will attempt to summarize what is involved in each proposal or idea; where, when and by which party the proposal was made; the course and extent of its development, if any; estimates, where they have been made, of potential revenues; and any comments made in the presentation and consideration of the proposal concerning its practicability. A summary evaluation will also be made noting those proposals which seem most significant as well as feasible, attainable or closest to realization.

Inventory

(a) International taxation of trade flows: revenue taxes

Although various studies have been made within the United Nations since its founding on the subject of international taxation, including effects of foreign trade, investments, double taxation and other matters, the subject of taxing international trade flows as a source of revenue was not presented until the United Nations Conference on Desertification. As noted previously, the Secretary-General of the Conference, in an informal note to the Committee of the Whole on 1 September 1977, suggested that in view of the substantial resources required to combat desertification, the competing demands on available resources and the need for financing on a long-term, sustained and predictable basis, the Conference should look into innovative financing mechanisms involving automaticity, including international taxation. The types of taxation discussed in sections A, B and C that follow are mainly of the revenue kind, i.e. taxes designed for the sole purpose of raising revenues, rather than as incentives to alter behaviour.

A *General trade tax*

As was noted previously, the first formal request by a United Nations entity for a study of a general international tax was made in the PACD, which suggested the creation of a *Special Account* that would draw revenues from various sources, 'including international taxation', and asked for a special study on fiscal measures entailing automaticity. Specific proposals for a general turnover tax, i.e. a tax on all trade transactions, were presented in the study contained in Part I of this book. The study noted that in 1978, total world trade should reach £1,000 billion, and that a tax of only 0.1 per cent of the value of international trade would yield £1 billion annually and would not, in itself, be a disincentive to world trade in view of its low rate.

The proposal for a general trade tax received an additional impetus from the Brandt Commission Report. The report noted that the volume of international trade had in 1979 reached £1,300 billion, and that a small surcharge on imports would not be difficult to administer. It estimated that a 0.5 per cent levy on international trade would yield about £7 billion annually.

B *Specific traded commodities*
1 OIL – CRUDE AND REFINED   In addition to suggestions relating to a general trade tax, proposals have been made for revenue taxes on

specific traded commodities. At the Thirty-First Session of the United Nations General Assembly in November 1976, the delegation of Saudi Arabia proposed a tax of $0.01 per barrel of oil (as well as a voluntary tax of 0.1 to 0.2 per cent on arms sales) to provide additional funds to meet global, environmental and other needs. The study on additional measures (Part I) also suggested a general tax on the trade of crude and refined oil products, on the grounds that it was the largest single component of world trade, with a value of approximately $130 billion in 1977. It was noted that such a tax might encounter some resistance from oil-exporting countries, but that some support could be forthcoming if some acceptable formula were developed. The Brandt Commission also included a tax on internationally traded crude oil among the recommendations in its report. At the Fourth Session in February 1980 of the Preparatory Committee on a New International Development Strategy, the ACC Task Force on Long-Term Development Objectives also suggested oil exports among the possibilities for international taxation.
2 OTHER – EXHAUSTIBLE MINERALS   In addition to crude and refined oil products, proposals have been made for taxing other specific internationally traded commodities. The Brandt Commission has proposed, as other specific traded commodities which can be taxed, hydrocarbons and other exhaustible minerals.

C *Invisibles*
In addition to taxing internationally traded commodities, proposals have been made for the taxation of 'invisibles'. The Brandt Commission has noted the possibility of levies on international investment, international air travel and freight transport. It estimated that a 1 per cent levy on international passenger and freight transport would yield $250 million per year, with a growth of 10 to 15 per cent per annum.

(b) Tax on reverse transfer of technology

Under one form, this tax would involve placing additional income taxes on migrants and remitting the proceeds of this tax to their countries of origin, at least until the cost of their education was covered. It is not, in this respect, strictly a revenue tax but also a compensatory tax for the reverse transfer of technology, or 'brain drain' from developing countries. These proposals were made by Professor J. N. Bhagwati for the Group of Governmental Experts on Reverse Transfer of Technology of UNCTAD in February 1978. At the International Labour Conference (Sixty-Third Session, Fourteenth Special Sitting) in June 1977, the Crown Prince Hassan bin Talal of Jordan also proposed the establishment of an 'International Labour Compensatory Facility' (ILCF) that would draw its resources principally from labour-importing countries, with the revenues diverted to developing labour-exporting countries in proportions related to the estimated cost of the loss of labour. The General Assembly, in resolution 34/200 of 19 December 1979, *inter alia*, requested the Secretary-General to carry out, in close co-operation with UNCTAD, the ILO and other relevant United Nations bodies, a study on the feasibility of the proposal for the establishment of an international labour compensatory facility and to submit a progress report to the General Assembly at its Thirty-Fifth Session and a final report at its Thirty-Sixth Session.

A third United Nations forum in which a 'brain drain' tax proposal was presented was at the United Nations Conference on Science and Technology held in Vienna in August 1979, which, at the initiative of the Group of 77, adopted as a part of the Vienna Programme of Action, a proposal for the mobilization of resources to be accrued from the ILCF. The Programme of Action was endorsed by the General Assembly in its resolution 34/218 of 23 January 1980.

Taxation of the reverse transfer of technology will also be among the possible means of financing to be considered in a study by an Intergovernmental Group of Experts which will prepare a Financing System for Science and Technology for Development, as requested in General Assembly resolution 34/218. The group, which will hold an initial meeting from 11 to 15 August 1980, will submit the study to the Intergovernmental Committee on Science and Technology for Development, that will consider the study and make appropriate recommendations to the General Assembly at its Thirty-Sixth Session for decision. It is envisaged that the Financing System will come into effect on 1 January 1982.[1]

(c) Tax on Surpluses in Balance of Trade

At the United Nations Conference on Science and Technology for Development, the Group of 77 also presented a proposal for contributions from developed countries which would be calculated on the basis of a percentage, to be determined at a later stage, of the average quinquennial surpluses of developed countries in their balance of trade in manufactured goods with the developing countries. The revenues generated would be used to help finance the Vienna Programme of Action of Science and Technology for Development. The proposal originated at the Second Latin American Regional Preparatory Meeting for the Conference on Science and Technology for Development, held from 29 November to 1 December 1978 at Montevideo, Uruguay, by the Economic Commission for Latin America (ECLA).

(d) Consumption taxes

According to this proposal, a tax would be levied on a limited number of goods, the possession of which indicated the attainment of a relatively high level of living by the purchasers, e.g. private aircraft and pleasure-boats, automobiles, television sets, refrigerators, washing machines and dishwashers. The contribution would be assessed at a low uniform rate of about 0.5 per cent on the purchase price of the item, and would be collected by the tax authorities of eachs country on their own responsibility. The revenues would be used for financing international development. The governments would be able to choose the particular uses and would be considered to have fulfilled their pledges by submitting each year the receipts to international development organizations chosen by them from a list adopted by the General Assembly.

This proposal was discussed extensively by the Committee for Development Planning as a 'World Solidarity Contribution' at its Sixth Session in January 1970 (ECOSOC, Committee for Development Planning, Report of the Sixth Session; official records; 49th Session, Supplement No. 7, E/4776). The Committee did not come to an agreement on the proposal and recommended that in view of the complexities of the problem, a study be undertaken as soon as possible of the feasibility of the proposal. To date, no such follow-up study has been undertaken. It may be noted, however, that the Brandt Commission has also presented a proposal for a tax on durable luxury goods.

(e) Income from the use of the international commons

Another potential major resource for financing multilateral programmes lies in revenues from the exploitation of the international commons, i.e. those physical attributes of the universe which are not subject to national sovereignty and may be considered the common heritage of all mankind. One can include under this category much of the world's oceans, including the marine environment, outer space, telecommunications channels, the moon and Antarctica as well as parts of the continental shelves.

A *Ocean resources*
The world's oceans cover 75 per cent of its entire surface and, as is well known, supply an enormous part of world's resources – both living and non-living.

1 LIVING Most of the world's oceans lie beyond national jurisdiction, and with their living resources as well as the right of passage, have been considered under International Law as the 'High Seas'. They are open, in principle, to use by all nations for their national interests, with no international economic obligations. The oceans' considerable fisheries resources have generally been excluded from discussions of the international commons, as has been the right of passage. Ideas have been put forward, however, for the charging of fees for the exploitation of the

oceans' living resources, as well as the right of passage. For instance, references have been made by the General Assembly (Thirtieth Session in 1975, by the delegate of Sri Lanka) and the Food and Agriculture Organisation (FAO) Council (Seventy-First Session in 1977, by the Committee on Fisheries) to the enormous unexploited living marine resources of the Southern Oceans and their possible exploitation and sharing to meet the world's future food needs. The Brandt Commission report also includes ocean fishing among the proposed automatic revenues for development.

2 NON-LIVING The subject of non-living ocean resources has been followed closely within the United Nations system since 1967, when the idea of the use of sea-bed resources 'in the interests of mankind' was first introduced at the General Assembly by the delegate of Malta. On 17 December 1970, the General Assembly, in resolution 2749(XXV), declared the non-living ocean resources beyond national jurisdiction to be the 'common heritage of mankind'. The possible agreement on a scheme for charging fees for income from the deep ocean bed is being discussed by the Third United Nations Conference on the Law of the Sea. [See *United Nations Convention on the Law of the Sea*, 1982.]

Among the proposals being considered is the establishment of a 'Common Heritage Fund', which had first been presented by the delegation of Nepal during the Seventh Session of the Conference in 1978. The Fund would derive its income from a share of net revenues from (a) mineral exploitation of the sea-bed and sub-soil of the exclusive economic zones; and (b) those portions of the continental margins beyond the exclusive economic zones outlined in the Convention. The revenues of the Fund would be used principally to assist developing countries and also to help protect the marine environment and assist in the transfer of marine technology.

Support has also been expressed, without specifying operational details, for charges for the exploitation of the deep ocean bed by the Task Force on Long-Term Development Objectives of the United Nations Administrative Committee on Co-ordination (ACC) at the Fourth Preparatory Committee for the New International Development Strategy and by the Brandt Commission.

It should be noted, however, that the extent of the oceans' non-living resources available for international exploitation has been reduced by the assertion by coastal States of exclusive economic rights over zones extending 200 miles or more from land, which would, in effect, appropriate about one-third of the ocean area containing most of the known hydrocarbon and other mineral resources of the sea. The only promising resources presently known to lie in the deep ocean bed beyond the exclusive economic zones claimed by coastal States are manganese nodules, which have some value for their manganese, nickel, copper and cobalt contents.

The UNEP expert group on additional measures recommended that economic rents be appropriated for international purposes on the mining of these manganese nodules. The group estimated, on the basis of a United Nations study on the subject, that a nodule mineral industry might produce economic rents (i.e. net revenue after costs and a normal return on capital) of several hundreds of millions of dollars annually. In a note to the Third Regular Session of 1979 of the Administrative Committee on Coordination (ACC/1979/72), the Director-General for Development and International Economic Cooperation pointed out that taxes on mining the sea-bed would not become available until the early 1990s, and that the magnitude of the income would be moderate at best.

B *The moon*

In resolution 34/68 of 5 December 1979, the General Assembly arrived at an Agreement Governing the Activities of States on the Moon and Other Celestial Bodies. The Agreement specified that 'the moon and its natural resources are the common heritage of mankind', that the exploration and use of the Moon shall be the problem of all mankind, carried out for the benefit of all countries irrespective of their degree of economic or scientific development, and that there should be an equitable sharing by all parties to the Agreement in the benefits whereby the interests of the developing countries as well as the efforts of those countries contributing to the exploration of the Moon shall be given special consideration. The Agreement also stated that due regard shall be paid to the interests of present and future generations as well as to the need to promote high standards of living and conditions of economic and social progress and development. Finally, the parties to the Agreement would undertake to establish an international regime to govern the exploration of the natural resources of the Moon as such exploration approached feasibility (article 11(5)). A conference is to be convened ten years after the entry into force of the Agreement, during which the establishment of the international regime will be considered. The Agreement, annexed to General Assembly resolution 34/68, will have to be subject to the usual ratification procedures before it enters into force.

C *Antarctica*

Resolution of the legal status of Antarctica has been deferred by international agreement, and the international community has been careful to avoid the question of the inclusion of the Antarctic continent and adjacent seas as part of the inclusion of the Antarctic continent and adjacent seas as part of the global commons. No proposals, for instance, have been made in this regard at the Third United Nations Conference on the Law of the Sea. It may be noted, however, that a number of relevant observations have been made in United Nations for regarding the resources of the continent. At the Thirtieth Session of the General Assembly in October 1975, for instance, the delegate of Sri Lanka noted that there were still areas where there could be constructive and peaceful co-operation on the part of the international community for the common good of all rather than for the benefit of the few, and that such an area was the Antarctic continent. At its Eleventh Session, held in June 1977, the Committee on Fisheries of FAO noted the various known and potential resources of Antarctica and the southern oceans – notably krill and hydrocarbons – which are believed to be substantial, and the need for carrying out development activities with due regard to the requirements of conservation in relation to the Antarctic ecosystem. It was agreed that FAO would keep the Antarctica Treaty countries informed of its relevant activities.

D *Telecommunications and satellites in geosynchronous orbits*
The UNEP expert group on additional measures noted in the study contained in Part I of this book that two special attributes of the physical universe can properly be regarded as part of the international commons: telecommunications wave-lengths, and parking spaces or 'slots' in outer space for satellites in geosynchronous orbit. Economic rents produced by exploitation of these physical attributes could, in principle, be appropriated for international purposes.

The use of telecommunications channels already produces significant rents, but appropriating them for international purposes would encounter serious complications. The assignment of frequencies is largely in the hands of national governments, and the profits of the firms using those frequencies are already subject to national taxation. Moreover, an increasing part of the long-distance use of telecommunications frequencies involves satellites belonging to INTELSAT, which already charges for the use of its circuits. Alternatively a small national surtax on all telephone and telegram/telex charges was suggested, the proceeds of which could be used for international purposes.

Charges on parking spaces or 'slots' for satellites in geosynchronous orbit may one day become a source of revenue for international purposes. These slots are rapidly becoming a scarce resource, and the time may come when international action will be needed to control the allocation of the choice slots that are well located with respect to economically-important land areas. A good case can then be made for charging for the use of such slots.

Proposals for charging rents for the use of telecommunications channels and parking spaces for satellites in geosynchronous orbits were also made by the Brandt Commission. The legal implications of satellites remaining in a stationary position relevant to a point above the earth are still under discussion and were on the agenda of the Legal Sub-Committee of the Committee on the Peaceful Uses of Outer Space which met in Geneva in March and April 1980. Some delegates pointed out that the geosynchronous orbit constituted a limited natural resource, over which the equatorial countries exercised sovereign rights. Other delegates maintained that geostationary orbits were inseparable from outer space which, under the terms of the Outer Space Treaty – including the geostationary orbit – were not subject to international application, and that the orbits were free for use by all States. Other States were of the view that a special regime for the geostationary orbit should be formulated.

E *Taxes on polluters of the marine environment*
This proposal, although called a 'tax' in the study contained in Part I, is intended not so much as a revenue-raising mechanism as a punitive/incentive measure to reduce marine pollution. One of the ocean's most visible pollutants of the marine environment are oil and petroleum products. According to the study, a tax on the amount of oil or refined products discharged into the ocean would induce polluters to reduce their polluting emissions and, to the extent that total elimination of emissions was uneconomic, produce some revenue for international purposes. It therefore proposed that polluters who installed anti-pollution technology be taxed at a lower rate than those who did not. For example, tankers with segregated ballast tanks do not, on the average, discharge as much oil into the oceans as those that are not so equipped. A differential tax on these two categories of tankers could induce owners to order new tankers with segregated ballast tanks when old tankers are retired. In the first year, such a tax might yield about $30 million. In the twentieth year, assuming that all tankers would by then have segregated ballast tanks, revenues would fall to about half that amount because of the lower tax rates on pollution by tankers with segregated ballast tanks. Average annual revenue over the 20-year period was estimated at more than $20 million.

In view of the particular severity of the problem in the Mediterranean, the 'polluter-pays principle' was raised at an Intergovernmental Consultation of Experts on a regional oil-combating centre convened by the Executive Director of UNEP in accordance with a recommendation by the Intergovernmental Meeting on the Protection of the Mediterranean in 1975.

(h) Savings from disarmament

Since the inception of the League of Nations, disarmament has been a subject of continuing international debate and aspiration. For more than 20 years, the United Nations has also discussed the subject extensively, with a view to promoting international peace and security and also to releasing funds for the development of the poorer countries of the world. The relationship of savings from disarmament to development has been discussed in different United Nations fora, particularly the General Assembly, where various proposals and resolutions have been introduced and adopted.

The principle that there should be a reduction in arms expenditures and the diversion of the expenditures for development has been presented in a number of General Assembly resolutions. These include resolution 380 (V) adopted in November 1950 which called for a reduction of arms expenditures to a minimum and the use of resources released for the general welfare, with due regard to the needs of the developing countries.[3]

At its Tenth Special Session held in May and June 1978, the General Assembly, *inter alia*, requested the Secretary-General, with the assistance of a group of qualified governmental experts appointed by him, to initiate an expert study on the relationship between disarmament and development. The Final Document pointed out that the idea of a link between disarmament and development was not new, and that the major reason why it had not been implemented was that there had been no objective or generally acceptable ways of measuring the true levels of armaments. It may be noted, in this connexion, that a detailed system and format for the international reporting of military expenditures for the General Assembly to consider and possibly introduce into general use, had been prepared by the Secretary-General, assisted by a Group of Experts on the Reduction of Military Budgets, appointed in accordance with General Assembly resolution 3463 (XXX) of December 1975. The report, 'Reduction of Military Budgets: Measurements and International Reporting of Military Expenditures' (A/31/222/Rev.1), was presented to the General Assembly at its Thirty-First Session in 1976. The UNEP expert group on additional measures (study contained in Part I) urged its adoption by the General Assembly, but the proposed format had not been approved thus far.

The proposals for channelling savings from disarmament to development have also been presented in other specific forms. General Assembly resolution 3093A (XXVIII), adopted in December 1973, was specifically directed at the States which are permanent members of the Security Council and suggested, *inter alia*, that they should reduce their military budgets by 10 per cent from the 1973 level during the next financial years and allot 10 per cent of the funds released for assistance to developing countries.

In the study *Disarmament and Development: An Analytical Survey and Pointers for Action*, prepared for the Committee for Development Planning in January 1977, it was estimated that a 10 per cent reduction in United States–USSR military expenditures alone would mean a 'disarmament dividend' of $23 billion a year. If this dividend were to be shared in the proportion of 90 per cent for social and economic programmes in the United States and the Soviet Union themselves, and 10 per cent for the provision of increased concessional capital to the poorest countries, there would be a substantial annual increase of $2 to $3 billion, representing a 20 per cent increase in official development assistance over the previous year.

The study contained in Part I estimated (for 1977) that the level of world military expenditure was on the order of magnitude of $360 billion annually. A reduction of merely 5 per cent could release some $18 billion a year. If one-half of this amount could be earmarked for international programmes, over $9 billion a year would be available for these purposes. An intensified effort to reverse the present trend in arms expenditure was imperative.

In the course of its activities on the subject of disarmament and development, the United Nations has commissioned studies and appointed expert groups to deal with the problem. The most recent body established is the Group of Governmental Experts on the Relationship between Disarmament and Development, which was established under the provisions of the Final Document of the Tenth Special Session, devoted to disarmament, of the General Assembly held in 1978. The Group commissioned 45 research projects in areas where disarmament could have a catalytic impact on development. The projects were to be submitted and reviewed in June 1980, so that the group could complete its final report in time for consideration by the Second Special Session of the General Assembly on disarmament which will be held in 1982.

[The Second Special Session of the General Assembly launched a World Disarmament Campaign under United Nations auspices in conformity with the principles established at the First Special Session held in 1978. The Campaign has three primary purposes: to inform, to educate and to generate public understanding and support for the objectives of the United Nations in the field of arms limitation and disarmament.]

(i) **Special drawing rights (SDR)–development link**

In the monetary field, proposals have been made for a link to development of the SDR, which were established and have been issued from time to time by the IMF since 1968. The SDR are essentially a permanent line of credit in the IMF from which member countries can draw under certain conditions, to obtain the foreign currencies they need. SDR have thus become an important reserve asset, along with gold and foreign currency reserves, in countries' Central Banks and Treasuries. Even before the SDR were established, proposals were made, in anticipation of the time when a new kind of reserve asset would have to be created, for a link of the new reserve asset to development purposes.

The proposals for a link to development were originally made by an UNCTAD Group of Experts in 1965, which prepared a report for the Trade and Development Board entitled 'International Monetary Issues and the Developing Countries'. Among the relevant conclusions of the UNCTAD Expert Group were that: (a) the general level of reserves was inadequate or about to become so; (b) developing countries had a legitimate and pressing need for additional liquidity, which should be provided in part through the expansion of reserves and by increasing the amount of credit facilities available from the IMF; (c) it was both feasible and desirable to establish a link between the creation of international liquidity and the provision of development finance without detriment to either process; and (d) the reform of the international monetary system should be truly international, with developing countries represented in the discussions leading up to monetary reform and in the operation of the new arrangement, in accordance with the degree of their interests and concerns.

When the IMF finally decided to increase international liquidity through the creation of SDR as a reserve asset, no special relationship was provided for a link to development. In fact, the SDR, which were created solely for increasing international liquidity, were allocated in proportion to the quotas of the members of the IMF. Thus, the richer countries (which had the largest quotas) obtained the greatest share of the SDR, while the poorer countries, with their low quotas, obtained only small amounts of the new reserve assets.

In view of the above, proposals have been made for a different system of distributing SDR, taking into account the countries' level of development and needs. No precise formulae have been presented in the United Nations fora for a new key for the distribution of SDR. The Group of 77, however, has repeatedly pressed for a different system of allocation of SDR in various United Nations fora, including the General Assembly and UNCTAD. At the IMF itself, where the necessary international monetary reforms giving effect to a link would be made, the Group of 24 has also made similar proposals. These are included in the Programme of Action on International Monetary Reform which was prepared by the Group of 24 and endorsed by the Group of 77 at a meeting in

Belgrade on 29 September 1979, and in a communique issued in Hamburg on 24 April 1980 prior to the annual meeting of the IMF Board of Governors. The case for the link has also been presented unsuccessfully in the deliberations of the Committee on Reform of the International Monetary System and Related Issues of the IMF Board of Governors – the 'Committee of Twenty' – which was created by the IMF to include representatives from both developed and developing countries.

On the means by which the link could be established, several variations have been proposed: (a) the channeling of SDR through an international financing organization, such as the International Development Association (IDA), as proposed, among others, by the UNCTAD Group of Experts; and (b) the creation of additional SDR and their direct distribution to members of the IMF, according to a key different from that based on the quotas of IMF members. This is the approach taken by the Group of 77.

It has been noted that the reforms proposed would probably involve amendments to the IMF's Articles of Agreement. Since SDR, under IMF's, rules, can be allocated in principle only to Central Banks and Treasuries, an amendment would be required to allow the allocation of SDR to an international financing organization such as IDA. One proposal that would circumvent the need for such an amendment would involve the creation of additional SDR with a tacit understanding that the developed countries would voluntarily contribute a proportion of their allocations to the international financing organization whenever new SDR are created. Such a proposal was made by I. G. Patel in a study for UNCTAD in 1967, 'The Link Between the Creation of International Liquidity and the Provision of Development Finance', (TD/B/115/Add.2 and TD/B/C.3/43/Add.2; 14/8/67). This approach was also suggested at the 1968 IMF annual meeting by Finance Minister Emilio Colombo of Italy (International Monetary Fund, *Summary Proceedings* of the twenty-third annual meeting of the Board of Governors, 1968, page 81). Most of the other proposals which have been presented, however, have been for a more binding agreement, through an amendment to the Articles of Agreement of the IMF.

No progress has been recently made on the establishment of an SDR-development link at the IMF. The idea of a link, however, is being kept alive by the Group of 77 at the United Nations and by its related Group of 24 at the IMF, as well as by the study contained in Part I. The idea of an SDR-development link has also received new impetus, along with other proposals, in the recommendations of the Brandt Commission report.

(j) Proceeds from IMF gold sales

Aside from the proposed SDR-development link, a major potential monetary resource for multilateral programmes lies in the large holdings of gold by the IMF. Proposals for the tapping of these resources have, in fact, been made and were realized with the Agreement adopted in August 1975 by the IMF's Interim Committee of the Board of Governors on the International Monetary System to reduce the role of gold in the International Monetary System while assisting the developing countries. The agreement provided for the abolishment of the official price of gold; it also provided that one-sixth of the IMF's gold holdings would be sold, with the profits going to the developing countries, and that another one-sixth would be returned to members.

At the Jamaica meeting in January 1976, the Interim Committee, to implement its earlier decisions, agreed that the profits would be channelled through a Trust Fund within the IMF. The Trust Fund was established in May 1976 for the purpose of providing additional balance-of-payments assistance on concessional terms to developing member countries. The resources available for loans by the Trust Fund are derived mainly from the profits of the sale, through auctions, of 25 million ounces of the Fund's gold.

Each disbursement under a Trust Fund loan is to be repaid in 10 equal semi-annual instalments, which begin not later than the end of the first six months of the sixth year after the date of the disbursement, and are to be completed at the end of the tenth year. Interest is charged at the rate of one-half of 1 per cent per annum on the outstanding balances of the loans and is paid in semi-annual instalments. Although these terms are not as concessionary as those governing IDA credits, they are highly favourable, especially in comparison with other terms for IMF loans, and in view of current high interest rates and rates of inflation.

The Trust Fund's operations have been divided into two periods, the first from 1 July 1976 through 30 June 1978. During this period, Trust Fund loans totalling the equivalent of SDR 841 million [SDR = $1.15 per SDR] were made to 43 IMF members. The second period ran from 1 July 1978 through 30 June 1980, when the Trust Fund ended as a source of assistance to developing countries. To date, loans amounting to SDR 800 million have been made during the second period.

Recently, two other proposals were presented for the use of proceeds from IMF gold sales for development purposes. At the Fourth Session of the Preparatory Committee for the New Development Strategy held in February 1980, the ACC's Task Force on Long-Term Development Objectives proposed the establishment of an International Development Fund that would, *inter alia*, be designated as a beneficiary of resources generated in the public domain, including profits from the sale of monetary gold.

The Brandt Commission has also proposed that the remaining IMF gold stocks, which amount to about 100 million ounces, be utilized for development purposes. This could be accomplished by using the gold stocks as collateral against which the IMF could borrow from the market, the borrowed funds then being made available to developing countries, particularly the middle-income groups. Further staggered sales of some of the gold could also be made and the profits from such sales used as interest subsidies on loans to low-income developing countries.

More generally, the Brandt Commission has also suggested that profits from gold sales be used for development assistance in ways to be decided by the members of the IMF. The Commission estimated, in this connexion, that profits from the progressive sale of the remaining two-thirds of the IMF's gold holdings – about 100 million ounces – could amount to a sum in the order of $30 to $40 billion over a period of time, assuming a market price of $300 to $400 per ounce of gold. Such a sum could yield annual revenues in the order of $2.4 to $3.2 billion, assuming a return of 8 per cent. Because of the fluctuating price of gold, the Brandt Commission's estimates are, of course, subject to change.

(k) Commodities Stabilization and the Common Fund

In the area of trade and development, multilateral compensatory measures for commodities exports fluctuations – which involve automatic mechanisms – have been the subject of discussion in various United Nations fora since the early 1950s. As was noted previously, the General Assembly, in resolution 623 (VII) of 21 December 1952, expressed its concern over the declining terms of trade of primary commodities in relationship to capital goods and other manufactured articles and, *inter alia*, requested the Secretary-General to appoint a Committee of Experts to consider the subject. The Committee's report pointed out the deficiencies of single commodity agreements and outlined schemes for automatic compensation of large short-term changes in the terms of trade.

The question of international compensation for fluctuations in commodity trade has since been discussed extensively by the General Assembly, the Economic and Social Council and its former Commission on International Commodity Trade, as well as by the FAO Committee on Commodity Problems. General Assembly resolutions 1324 (XIII) of 14 December 1953 and 1423 (XIV) of 5 December 1959, as well as ECOSOC resolution 726 (XXVIII) of 24 July 1959 called for proposals for automatic compensatory financing of fluctuations in commodity trade. These were also discussed extensively by the Commission on International Trade and the FAO Committee on Commodity Problems. In response to requests by these bodies, a number of reports were prepared containing proposals for compensatory mechanisms.[4]

The major breakthrough on automatic compensatory mechanisms was reached by UNCTAD IV, which met in Nairobi in 1976 and adopted an Integrated Programme for Commodities with a time-table for the negotiation of the Programme, comprising mainly a Common Fund and International Commodity Agreements (ICA). Since then, intensive negotiations have been carried out on the Common Fund and ICA. The Common Fund would derive its resources from an initial paid-in capital for supporting stock operations, from borrowing and from cash deposits and guarantees from ICA associated with it. The financing arrangements for the Common Fund thus involve a measure of automaticity.

While the initial discussions concerned mainly the financing of buffer stocks to stabilize commodity prices, the Agreement recently reached for the establishment of the Common Fund provides for a Second Account or 'Window' which would finance research and development, productivity improvements, marketing and other development measures for commodities in developing countries. Although the resources would come mainly from voluntary contributions, earnings derived from the First Account may be allocated to the Second Account. In certain cases funds made available through the Second Window for developmental measures in respect of individual commodities produced in countries where desertification is a problem might provide support for anti-desertification efforts.

(l) Other fiscal measures involving automaticity

Another proposal for the automatic generation of resources has been the establishment of a world-wide lottery under the sponsorship of the United Nations. This proposal was presented by the delegation of Ghana to the Administrative and Budgetary (Fifth) Committee of the General Assembly, during its Twenty-Sixth and Twenty-Seventh Sessions. The Committee decided to request the Secretary-General 'to undertake an evaluation of a world-wide United Nations Lottery as a potential source of revenues'. The Secretary-General, in a report based on responses to a questionnaire sent to Member States, concluded that the General Assembly would not wish him, for the time being, to engage the services of an outside consultant to study the question. The Fifth Committee concurred with this conclusion, although it was pointed out that not enough replies were made to the questionnaire.[5]

Other than the above, the only other measure involving an element of automaticity is a 'tax-like' proposal for the introduction of special contributions by multi-national corporations operating in developing countries. This proposal was presented at the Sixth Session of the Committee for Development Planning in 1970, but no decision was taken on it.

Related considerations concerning automaticity
In the presentation of proposals for the generation of resources for multilateral programmes involving automaticity, the related elements of the mechanism for the mobilization or the resources, burden sharing and the allocation of the proceeds have also been dealt with, in varying degrees of detail.

Mobilization
The study contained in Part I acknowledged that this was a complicated problem and made a number of observations. It also noted that in some cases the collection of the new sources of revenue could be assigned to existing mechanisms. National customs officials could collect international taxes on trade, and national port authorities could enforce pollution charges on shipping. In other cases, new mechanisms would be necessary. For example, the proposed International Sea-Bed Authority would seem to be the proper agency for collecting revenues from the exploitation of non-living resources of the ocean beyond national jurisdiction.

Burden-sharing
On the question of burden-sharing, it is acknowledged that the proposed sources of financing would not impose equal burdens on all countries, either in total magnitude or in relationship to their GNPs. For instance, international trade taxes would impose an especially heavy burden on international trade, and adjustments would have to be made to ease that burden.

The UNEP expert group (Part I of this book) stated that this fact was not in itself a reason for rejecting international taxes, since differential burdens are an unavoidable attribute of all workable revenue systems, both national and international; nevertheless, adjustments of national burdens would clearly be required in the interests of equity and political acceptability. Thus, the burdens imposed on individual countries should, in principle, be directly related to their level of development. This result could be achieved by permitting poor countries to retain part or all of the funds they collect for international purposes (e.g. in the case of trade taxes), or by selecting sources of revenue which, by their nature, benefit the poorer countries rather than the rich (e.g. the SDR-development link). The distribution of benefits from the disbursement of international revenues should also be taken into account in evaluating the fairness of measures used to create those revenues.

The question of a more equitable sharing of the costs of financing the United Nations system's operational activities is also touched upon in a report by the Director-General for Development and International Economic Co-operation, 'Some Policy Issues Pertaining to Operational Activities for Development of the United Nations System' (Document A/224/Add.1), for the General Assembly at its Thirty-Fifth Session. The report also discusses other aspects concerning the need to generate resources available for operational activities on a predictable, continuous and assured basis.

The Brandt Commission report, in reference to this problem, emphasized that international taxation must be universal, and that all countries must share the burden. It noted that universality calls for contributions not only from the industrialized Western countries, but also from Eastern countries and from developing countries, except the poorest.

Allocation of proceeds
As the principle of automaticity in the mobilization of resources is a relatively recent one, the suggestions made thus far for their allocation have been limited. The study contained in Part I suggested that those funds raised or contributed directly for anti-desertification programmes should be the responsibility of the UNEP Governing Council, which is the body entrusted by the General Assembly with the follow-up and co-ordination of the implementation of the Plan of Action to Combat Desertification. It also suggested that the Consultative Group for Desertification Control (DESCON), and a public international corporation (see Part III Section V of this book) proposed for the mobilization of funds for anti-desertification projects, should be used, if established, in the financing of anti-desertification programmes and projects.

Other proposed mechanisms for the allocation of automatically-generated resources for specific purposes include the International Sea-Bed Authority and its related Enterprise, which would be established under the United Nations Convention on the Law of the Sea. In the case of IMF gold sales, the IMF itself administered the distribution of profits in the form of loans to developing countries through the establishment of a trust fund.

In the case of the proposed SDR-development link, the IDA or some other intergovernmental financing agency has been proposed as the mechanism for the allocation of SDR. No precise formula has been presented in the case of the direct distribution of SDR to developing countries.

In the Brandt Commission report, a case has been made for the creation of a new institution which might be called a 'World Development Fund'; it would not be an alternative to existing institutions, but rather a supplementary institution that would channel resources mobilized through automatic means. The proposed institution would be established by all countries – 'West and East, and South excepting the poorest countries'. It would complement the lending of the World Bank and the IMF and co-operate with them as well as the regional development banks in financing development.

Evaluation
It is evident from this inventory of new means of financing international needs proposed in various fora of the United Nations that extensive consideration has been given to the principle of automaticity, and that this principle is gaining momentum within the United Nations system. Many of the automatic means of financing cited in this inventory could be an effective means of augmenting the resources of developing countries without straining the national budgets of the developed countries. The provision of increased assistance on an assured and more predictable basis is essential for development. This is especially so in developing countries suffering from desertification, for which programmes of a sustained and long-term nature are urgently required. Although there will be political opposition to many of the proposals which have been presented, it should be noted that automatic mechanisms have a clear advantage over the present system of voluntary contributions in that they would resolve the problems that governments and bilateral assistance programmes often have to face in trying to obtain the local political support necessary to increase the level of assistance – which indeed they must – to developing countries.

Among the large number of proposals listed in this inventory, several appear to be feasible, practicable or attainable in the near future. These measures could raise funds for any agreed international purpose, including combating desertification.

First, it should be noted that the system of gold sales by the IMF and the use of the profits for assistance through a trust fund has, by its very establishment and operation, proven its feasibility, practicability and desirability. This experience suggests that the IMF may again decide to resume the sale of its gold stock – presently about 100 million ounces – and revive its trust fund operations. On the basis of a market price of $400 to $600 per ounce, the profits from the sale of the IMF's gold holdings could yield a sum in the order of $40 to $60 billion over a period of time. Another measure could be to use the IMF's gold holdings as collateral for mobilizing funds in financial markets for re-lending to developing countries. In considering the eligibility of countries for loans or grants from a revived trust fund, desertification should be recognized as an important factor affecting the countries' balance-of-payments problems.

There is evident merit in pursuing the long-standing proposals for an SDR-development link, through the issuance of additional SDR and their distribution to developing countries either directly using a new key or through an international financing institution. The SDR have become an important part of the reserves of countries' central banks and treasuries, and developing countries have a need for a much greater share than they receive under the present system of distribution. The group does not consider that the establishment of a development link would lessen the standing of SDR. Like gold sales, the generating of resources for development through a link with SDR could play a significant role in accelerating development.

Another multilateral mechanism involving automaticity is the Integrated Programme for Commodities, including International Commodity Agreements (ICA) and the Common Fund. While this multilateral programme involving automaticity is close to realization, it should be noted that the financial resources involved are likely to be relatively small. This is especially so as concerns the amount of funds generated on an automatic basis that could be devoted to the Fund's Second Window. This programme can therefore be regarded only as a starting point.

Most economically exploitable non-living ocean resources appear to lie within the exclusive economic zones claimed by coastal States. The United Nations Convention on the Law of the Sea Conference legitimates these claims, and approves obtaining revenues for international proposals from the sea-bed beyond the exclusive economic zones and from the parts of the zones lying on the continental shelf more than 200 miles from shore. Modest international revenues may then be obtainable from mining manganese nodules and possibly from extracting oil and gas from the designated parts of the continental shelf.

The proposal involving automaticity with perhaps the greatest potential is an international trade tax. In view of the enormous volume of world trade – $1,300 billion in 1979 – a tax could be levied at a rate which would generate substantial revenues without acting as a disincentive to trade expansion, although there can be arguments concerning questions of its economic repercussions and equity. The international taxation system should include provisions to ensure that the poorest countries are not hurt. Acceptable alternatives could be taxes on specific commodities, such as oil, uranium and bauxite.

Another proposal involving automaticity which may appear attainable is the proposal to charge fees for 'parking slots' of satellites in geostationary orbit. No claims have been legally established so far in this part of the international commons, which is presently not subject to regulation. It is suggested that with the increase in the number of satellites in geosynchronous orbit, a system of regulation accompanied by charges which could be determined by auction might be established by the international community.

In terms of timeliness, practicability and feasibility the proposals reviewed in the foregoing four paragraphs appear to be the ones most likely to be attainable in the near future. That is not to say, however, that other proposals cited in the inventory are lacking in merit or unworthy of further study. Some at least are undoubtedly destined in due course to come forward as candidates for more intensive consideration; and it is entirely possible that a number may, at a later date, find acceptance.

The proposal for taxing military expenditures and the linking of disarmament to development could, if implemented, release substantial resources for the benefit of developing countries. In view of the state of international politics, however, it does not seem realistic to expect agreement on the proposals in this area at the present time or in the near future.

In the final analysis the prospects for adoption of any of the proposals mentioned depend not so much on technical feasibility as on the political will of member States. The increased international resources that would be generated by automatic means, furthermore, would be available for anti-desertification programmes only to the extent that it is so provided under a system of allocation. The first prerequisite to constructive action is that member States should understand and accept the fact that the problem of desertification vitally affects the lives and well-being of hundreds of millions of the world's inhabitants – and that if its ravages are left unchecked, it will surely pose a critical threat to tens of millions more. But understanding and acceptance alone are not enough: the second pre-requisite is that governments, through the United Nations, manifest their willingness to take the political action required if the challenge posed by desertification is to be met.

III FINANCIAL PLAN FOR COMBATING DESERTIFICATION

The PACD comprises 28 recommendations, all of which must be carried out to achieve the goal of stopping further desertification. The cost of many of the actions called for by these recommendations cannot, however, be estimated with any confidence at this time. The financial plan presented here therefore deals principally with those aspects of the Plan of Action the costs of which can be assigned at least rough orders of magnitude. Execution of only these aspects of the Plan of Action, it must be emphasized, would not stop further desertification. All other parts of the Plan would also have to be carried out.

Magnitude of the problem

About 47 million square kilometres, or one-third of the earth's land surface,[6] are arid or semi-arid. Roughly 20 per cent of these lands are hyperarid climatic deserts where agriculture is sustained in river basins and oases; the rest are areas with limited and irregular rainfall, used mainly for grazing and dryland farming. Between 600 and 700 million people depend on these fragile ecosystems for their livelihood.[7]

Because of the overuse and misuse of the land in these fragile ecosystems, extensive areas have been degraded, in the extreme to non-productive, desert-like conditions. The process of desertification, accompanied by declining yields, is continuing in the face of rising global food needs, which will require a 50 per cent increase in food production by the year 2000 if only to meet the present inadequate intake levels.[8] The total area affected by desertification is estimated at 3.3 billion hectares, or 80 per cent of the world's agricultural land in arid and semi-arid regions. A summary classification of lands affected, by type of land use, is shown in the following table:[9]

Type of Land Use	Total Area (000) ha	Area Affected By Desertification (000) ha	Percentage of Total Area
Irrigated land:	126,283	27,053	21
Rangeland:	3,751,071	3,071,603	82
Rainfed cropland:	224,428	173,127	77
Total:	4,101,782	3,271,783	80

The process of desertification results in sharply lower productivity than in lands not affected by desertification. This is reflected in losses of potential production and reduced incomes for the populations involved.[10]

The seriousness of the problem is also reflected in the amount of productive land deteriorating each year. According to preliminary estimates prepared for the UNCOD (1977), the equivalent of roughly six million hectares of productive land is going out of production each year. Another, more comprehensive measure of the damage caused by desertification is the area of productive land deteriorating each year to a level at which it yields zero or negative net economic returns. The annual rate of land degradation, following this definition, involves 20 million hectares, broken down as follows:[11]

Type of Land Use	Estimated area deteriorating annually to a level of zero or negative net returns (000) ha
Irrigated land:	546
Rangeland:	17,700
Rainfed cropland:	2,000
Total:	20,246

In certain cases, especially in developed countries, the farmers and ranchers or pastoralists are able to subsist and maintain their land-use systems through supplementary incomes. In most cases, however, the degradation of the land is reflected in malnutrition and poverty.

Scope and objectives of a financial plan

In paragraph 7 (b) of resolution 34/184, the General Assembly has asked for 'a financial plan and analysis outlining the components and costs of a programme to stop further desertification and identifying what is already being financed and what additional resources may be needed to meet the minimum objectives of stopping the spread of desertification'. The financial plan which is presented here is formulated within the framework of the PACD. The immediate goal of the PACD is to prevent and arrest the advance of desertification and, where possible, to reclaim desertified lands for production.

This financial plan does not cover the opening of new land for productive use, but includes corrective measures involving the improvement and restoration of desertified lands. Desertification is a continuing process involving the degradation of the productive capability of affected land. For purposes of formulating the financial plan, stopping desertification has been taken to mean preventing a further net loss of productive capability. This requires arresting the process of degradation on some land, reversing it on some land and, for economic reasons, accepting the fact of its continuance on some land.

The plan will comprise the following five components:

(a) corrective measures for irrigated land deterioration;
(b) corrective measures for rangeland deterioration;
(c) corrective measures for rainfed cropland degradation;
(d) the stabilization of moving sand dunes in critical areas;
(e) other essential measures whose cost cannot now be estimated.

In dealing with the first three, attempts will be made to estimate the extent of the areas affected, to examine the manifestations of the problem as well as their causes and solutions, and to calculate the cost of corrective measures. The land areas considered will include only those in arid and semi-arid regions.[12]

In line with the long-term goal of the PACD to implement it by the year 2000, the financial plan will be placed in a 20-year time-frame. It will deal only with the desertification that is taking place in irrigated lands, rangelands and rainfed croplands. While the areas referred to will include all such lands affected by desertification, the plan will deal only with the problems of desertification in developing countries requiring external financial assistance. For purposes of this study, per capita GNP of $1,000 has been used as the dividing line between countries requiring financial assistance and those that do not. Exceptions have been made for a few countries with per capital GNPs close to the line that are experiencing substantial favourable or unfavourable balances of payments. It is recognized that some countries not requiring financial assistance should be able to obtain technical assistance from United Nations agencies.

The plan will not cover the loss of land productivity resulting from mining, tourist development, development of highways and the expansion of urban communities, although it is recognized that these account for a significant amount of agriculturally productive land lost and other forms of environmental degradation each year.[13]

The component of the plan concerning sand dunes will cover the critical areas affected also in arid and semi-arid areas. It will not deal with the overall problem of moving sands, although it is recognized that moving sands are a serious problem in many countries.

The section of the plan on other measures that are necessary to implement the Plan of Action does not include all such measures, but only some of them that may be especially significant financially. Measures discussed include ecosystem conservation and the provision of alternative or supplementary livelihood systems and sources of energy to the populations involved, where overgrazing and the cutting of fuel-wood are a cause of desertification. Monitoring programmes for desertification control and strengthening government machineries for planning and implementation are also given attention.

The figures presented in the financial plan are necessarily only general orders of magnitude. High, medium and low illustrative figures have been calculated for most components of the plan. Estimates of unit costs of corrective measure are on the conservative (low) side in all three cases. More accurate and comprehensive estimates of the cost of carrying out the PACD must await the development of more detailed national programmes.

Corrective measures for irrigated land deterioration

The total area of irrigated land affected by desertification is the smallest among the three types of land use. It is also the most costly and most profitable to recover. The desertification of irrigated land is manifested principally in waterlogging, salinization and, to a minor extent, alkalinization. As the water table rises, bringing the salt contents up with it, productivity is reduced. Desertification results mainly from a lack of drainage, and/or use of low quality water. This is especially widespread in the world's arid and semi-arid areas.

The recovery of desertified land under irrigation is accomplished through the installation of proper drainage systems, leaching the soil of its salt content and, in the few cases of alkalinization, applying soil amendments. For the purposes of the calculations made in this study, the financial plan calls for restoring all irrigated land in developing countries requiring external financial assistance to a condition of optimum productivity.[14]

The average cost of restoring irrigated land to a condition of optimum productivity is estimated at from $500 to $1,200 per hectare, with an average which may be placed at $750 per hectare.[15] The cost of 'restoring' such desertified land should be distinguished from the cost of land 'reclamation' through the establishment or expansion of irrigation systems, which actually involves opening new land and is much more expensive.

Corrective measures to restore all arid and semi-arid irrigated lands in the world affected by desertification would have a 'medium' cost in the neighbourhood of $20.5 billion. This is based on the estimate that 27 million hectares are affected by desertification. Over 16 million hectares of the total area of desertified land under irrigation are in developing countries requiring external financial assistance, and the total cost of a programme for rehabilitating irrigated lands in those countries would range from $8.2 billion to $19.6 billion, with a medium estimate of $12.3 billion. On the assumption that this programme will be carried out over a 20-year period, which would be in accordance with the goal of implementing the PACD by the year 2000, the average annual expenditures would range from a low of $400 million to a high of $1 billion and a medium estimate of $600 million, in 1980 dollars. These estimates are detailed in Tables 3 and 4 (page 68).

Corrective measures for rangeland deterioration

As can be seen from Tables 1 and 2 (pages 63–66), the total areas of rangelands and of rangelands affected by desertification are the most extensive of the three types of land-use systems. It is estimated that over

Table 1 Arid and Semi-Arid Areas Affected by Desertification By Type of Land Use, Country and Region

	Irrigated land (000 ha)		Rangeland (000 ha)		Rainfed cropland (000 ha)	
	Total	Area affected by desertification	Total	Area affected by desertification	Total	Area affected by desertification
I Western Hemisphere						
North America:						
United States of America	15,500	1,650	235,000	188,000	30,000	15,000
Canada	300	60	10,000	7,000	5,000	3,000
Mexico	3,750	1,125	100,000	96,000	7,500	6,700
South America:						
Argentina	1,550	310	180,000	126,000	5,000	3,800
Bolivia*	65	6	12,000	11,500	1,000	950
Brazil	520	78	140,000	135,000	6,000	5,000
Chile	1,280	320	24,000	22,400	1,400	1,350
Colombia	0	0	3,500	3,200	0	0
Ecuador*	460	115	300	280	40	39
Paraguay*	9	2	12,000	9,600	50	20
Peru*	1,155	346	9,500	8,800	500	450
Venezuela	350	52	2,800	2,600	300	250
II Africa						
North Africa:						
Algeria	292	65	83,000	76,500	4,500	4,000
Egypt*	2,846	735	10,000	9,700	5	1
Libya	154	12	35,200	33,600	2,500	2,000
Morocco*	630	125	28,000	27,200	7,000	5,600
Tunisia*	128	50	10,100	9,000	3,000	2,100

Table 1 continued

	Irrigated land (000 ha)		Rangeland (000 ha)		Rainfed cropland (000 ha)	
	Total	Area affected by desertification	Total	Area affected by desertification	Total	Area affected by desertification
Sudano-Sahelian Region:						
Cape Verde*	1	0.03	0	0	45	30
Chad*	3.5	0.17	97,000	96,000	1,800	1,700
Djibouti*	0	0	2.2	2.2	0	0
Ethiopia*	30	5	85,125	77,000	3,500	3,100
Gambia*	29.5	5	0	0	200	100
Kenya*	20	1.1	22,000	21,000	300	270
Mali*	120	12	108,000	106,000	2,000	1,500
Mauritania*	28	0.15	72,300	71,000	150	100
Niger*	16.5	0.83	104,000	101,000	4,000	3,000
Nigeria*	13	0.6	30,000	28,000	5,500	5,200
Senegal*	144	0.10	13,000	12,100	2,400	2,000
Somalia*	100	9	63,600	57,500	1,000	950
Sudan*	1,610	250	203,000	198,000	3,500	2,600
Uganda*	4	0.2	375	350	0	0
United Republic of Cameroon*	2	0.3	10	6	8	7
Upper Volta*	5.0	0	16,000	15,500	2,700	2,500
Other countries south of the Equator:						
Botswana*	2	0.2	50,000	10,000	30	20
Madagascar*	670	45	5,000	4,800	200	150
Namibia*	8	0.1	66,000	16,500	10	5
South Africa	860	46	45,000	38,000	1,000	650
Tanzania*	40	4	28,000	14,000	2,400	1,900

III	*Zimbabwe**	0	0	7,500	4,000	300	150
	Australia	1,600	160	550,000	330,000	2,000	1,500
IV	*Asia*						
	West Asia/Middle East:						
	Bahrain	1	0	65	65	0	0
	Iraq	4,000	2,400	33,000	31,000	5,000	4,500
	Israel	170	17	1,240	1,100	190	185
	Jordan*	60	9	8,400	8,200	1,000	950
	Kuwait	1	0	1,580	1,500	0	0
	Oman	35	15	21,238	20,000	0	0
	Qatar	0.12	0	2,070	2,000	0	0
	Saudi Arabia	130	26	194,000	175,000	900	800
	Syria*	600	150	13,000	12,000	5,000	4,000
	Turkey*	2,000	600	11,000	9,400	9,000	6,000
	United Arab Emirates	5	0	8,366	7,900	0	0
	Dem. Rep. of Yemen*	5	0.5	28,500	22,800	0	0
	Yemen Arb Rep.*	100	20	18,300	17,000	600	350
	South-West Asia:						
	Afghanistan*	2,900	600	39,000	35,000	5,000	4,500
	India*	30,000	4,500	18,000	15,000	40,000	34,000
	Iran	5,250	1,320	127,000	124,000	6,000	5,500
	Pakistan*	13,300	5,050	56,000	54,000	6,000	5,400
	China*	19,000	3,700	370,000	307,000	8,000	7,400
	Mongolia*	30	4.5	93,000	56,000	900	150
IV	*Europe*						
	Spain	2,400	890	16,000	15,500	5,000	4,200
VI	Union of Soviet Socialist Republics	12,000	2,160	230,000	190,000	25,000	17,500
	GRAND TOTAL	126,282.62	27,052.78	3,751,071.2	3,071,603.2	224,428	173,127

Source: Professor Harold E. Dregne, Director, International Center for Arid and Semi-arid Land Studies (ICASALS), Texas Technical University.

* Developing countries requiring external financial assistance.

Table 2 Arid and Semi-Arid Areas Affected by Desertification, by Type of Land Use, showing (a) Areas in Countries not requiring Assistance; (b) Areas in Developing Countries requiring Assistance; (c) Areas in (b) to be included in Programme of Corrective Measures

Type of Land	Area affected by desertification (mil. ha)	Area of desertified land in countries not requiring external financial assistance (mil. ha)	Area of desertified land in developing countries requiring external financial assistance (mil. ha)	Area of desertified land in developing countries requiring assistance included in programme of corrective measures (mil. ha)
(1)	(2)	(3)	(4)	(5)
Irrigated land	27.05	10.71	16.35	16.35
Rangeland	3,071.60	1,626.36	1,445.24	722.62
Rainfed cropland	173.13	75.94	97.19	68.03
Total	3,271.78	1,713.01	1,558.78	807.00

Source: The figures given in this table are derived from data in Table 1.

82 per cent, or 3.1 billion hectares, of the world's arid and semi-arid rangelands are affected by desertification. This is due to the degradation of grazing lands over the years, the expansion of cultivated lands so that the worst lands have been left for grazing purposes, and the low inherent productivity of rangelands, especially in arid and semi-arid areas.

Desertification is manifested in the thinning or disappearance of vegetation, especially of fodder species. The natural vegetative cover is replaced by hardy, unpalatable species, or by short-lived vegetation. The main causes of rangeland deterioration are overgrazing and, in some areas, deforestation resulting from the cutting of fuel-wood. The salvaging of rangelands and the restoration of the range ecosystem to equilibrium and optimum productivity can be achieved through sound range management, including reducing livestock numbers, controlling the cutting of fuel-wood, and afforestation. There are considerable related additional costs in such measures, such as giving the people concerned alternative or supplementary livelihood systems and sources of energy (including planting of woodlots) to relieve the pressure on the land.

For the present, the cost of restoring rangelands is based on the immediate costs of better rangeland management. This involves grazing patterns that allow vegetation to recuperate. As a first step, the carrying capacity and stocking rates of the rangelands should be determined. Rotational grazing and the establishment of reserves in the event of drought are also used. Watering points are planned to give adequate access to all land being grazed. Reseeding and even fencing and guarding of grazing lands are necessary in some cases. The wholehearted cooperation of the people directly involved is essential to the success of these measures.

The direct costs of these aspects of proper range management are estimated at from $5 to as high as $400 per hectare where fencing and guarding are involved. The average low, medium and high costs per hectare have been estimated at $10, $20 and $40 respectively.[16]

The total cost of stopping the degradation of the world's rangelands affected by desertification, 3.1 billion hectares, using the estimated medium cost per hectare, would be on the order of $78 billion. A programme of this magnitude would clearly be unrealistic. The costs and areas involved are so vast that it does not seem feasible to restore the entire area even over a 20-year period. There are further considerations. It is estimated that only about 25 per cent of the rangelands in the arid and semi-arid countries can be recovered with positive economic returns.[17]

If the programmes were to try to restore only the approximately 25 per cent of rangelands affected by desertification which would yield positive benefit cost ratios, then the medium total cost would be in the

Table 3 *Alternative Estimates of Total Costs of a Twenty-Year Programme of Basic Corrective Measures for the Rehabilitation of Desertified Lands in Developing Countries Requiring External Financial Assistance*

Type of Land	Area Covered[1] (mil. ha)	Total Cost of Programme of Corrective Measures[2] (mil. $)		
		Low	Medium	High
(1)	(2)	(3)	(4)	(5)
Irrigated land	16.35	8,175.00	12,262.50	19,620.00
Rangeland	722.62	7,226.20	18,065.50	28,904.80
Rainfed cropland	68.03	13,606.00	17,007.50	30,613.50
Sand dune stabilization	2.04	449.00	449.00	449.00
Total	809.04	29,456.20	47,784.50	79,587.30

[1] *Source*: Table 2, column 5.
[2] Estimates are based on area covered and average low, medium and high unit costs of rehabilitation of desertified lands, as discussed in the text. Costings are based on the assumption that the affected areas in column 2 are moderately desertified.

order of $19.4 billion for all countries and $9 billion in developing countries requiring external financial assistance. This however, would not be an acceptable solution in view of the considerable populations whose livelihoods are directly at stake.[18]

If one included an additional 25 per cent of the affected rangelands in developing countries requiring external financial assistance, the total average annual costs would be of the order of $361 million (low), $903 million (medium) and $1.4 billion (high). These estimates are detailed in Table 4 (page 68).

Table 4 *Alternative Estimates of Average Annual Costs in a Twenty-Year Programme of Basic Corrective Measures for the Rehabilitation of Desertified Lands in Developing Countries Requiring External Financial Assistance*

Type of Land	Area Covered[1] (mil. ha)	Average Annual Cost of Programme of Corrective Measures[2] (mil. $)		
		Low	Medium	High
(1)	(2)	(3)	(4)	(5)
Irrigated land	16.35	408.75	613.13	981.00
Rangeland	722.62	361.31	903.27	1,445.24
Rainfed cropland	68.03	680.30	850.37	1,530.68
Sand dune stabilization	2.04	22.45	22.45	22.45
Total	809.04	1,472.81	2,389.22	3,979.37

[1] Source: Table 2, column 5.
[2] Estimated annual costs derived from Table 3.

The financial plan makes no provision for the remaining 50 per cent, which should be a question of national policy. Minimal measures involving no sructural improvements but including more efficient management might be applied to slow down or even arrest the rate of desertification of this land. Complementary measures such as supplemental or alternative means of livelihood could also be applied. Finally, uneconomic rainfed cropland can be converted to rangelands and fodder can be grown intensively around the perimeters of irrigated land to compensate for the decrease in livestock production in arid and semi-arid areas. To meet the world's increasing food requirements, the development of other promising lands should also be pursued.[19]

Corrective measures for rainfed cropland degradation

The deterioration of rainfed croplands is manifested in three ways: (a) depletion of soil fertility; (b) water and wind erosion; and (c) soil compaction and crusting. Depletion is caused by the leaching of nutrients, and erosion by the lack of protection from rain and wind. To restore rainfed croplands to a state of optimum productivity requires various measures. Where the problem is solely that of soil nutrient depletion, the solution lies in fertilization. Where the problem is erosion, compaction, or crusting, various soil conservation techniques are used, such as windbreaks, terracing, contour ploughing, strip cropping, water control and residue management.

Taken as a whole, the average costs of rehabilitating desertified rainfed croplands are estimated at $200 per hectare (low), $250 (medium) and $450 (high). Where the problem is solely that of soil depletion, the average cost of corrective measures is estimated at $50 per hectare. Where the problem is also that of erosion, the average cost is estimated at $500 per hectare.[20]

Approximately 173 million hectares of the world's total rainfed croplands in arid and semi-arid areas are affected by desertification, of which about 70 million hectares are subject to depletion of soil fertility and 103 million to water and wind erosion. Of the total, about 97 million hectares are in developing countries requiring external financial assistance. It is estimated that 70 per cent, or 68 million hectares, of these lands can be recovered with positive net economic returns. This is the area included in the financial plan. In the remaining 30 per cent, alternative or supplementary livelihood systems can be introduced for the populations involved (as in the case of rangelands), and a large part of the areas can be converted to grazing lands. The declining yields resulting from the desertification of rainfed croplands and their conversion to grazing lands could be compensated for by the development of other more promising lands to meet the world's growing needs, as detailed in the programme proposed by FAO in *Agriculture: Toward 2000*.

The medium cost of rehabilitating 70 per cent of the rainfed croplands affected by desertification in developing countries requiring external financial assistance would be of the order of $17 billion over a 20-year period. The estimated annual costs would be of the order of $680 million (low), $850 million (medium) and $1.5 billion (high). These estimates are detailed in Tables 3 and 4.

Stopping the encroachment of moving sand dunes

A special kind of desertification is in the form of the encroachment by moving sand dunes on productive agricultural areas, communities, oases and even roads and rivers. The problem of moving sands is a special one in that while the sandy areas themselves are already desert or desert-like (climatic or man-made), they pose a threat to productive and inhabited areas. In addition, the erosion of certain drylands has reached the point where sand dunes have formed and should be stabilized.

While the problems of moving sands is an extensive one, the financial plan includes only the critical areas affected. These include:[21]

(a) Sand dune fixation in hyperarid areas through non-vegetative means. The areas threatening oases, roads, airfields and small communities in hyperarid regions total roughly 78,000 hectares. This is the most expensive method of sand dune fixation and includes the use of bituminous and chemical covers as well as fences. The cost of fixation per hectare is estimated roughly at $1,000 based on samples of actual projects. The total cost of sand dune stabilization to protect threatened critical areas through non-vegetative methods would be of the order of $78 million.

(b) Green belts to protect cultivated lands and communities. The areas of moving sand dunes threatening cultivated lands and communities are estimated at 200,000 hectares. The average cost per hectare of establishing green belts is estimated at $300, giving $60 million as the total cost of this type of sand dune fixation.

(c) Revegetation for stabilization and range forage production. This is the most extensive form of sand dune fixation, involving a total area of 3.8 million hectares. The cost of revegetation is estimated at $200 per hectare, slightly less than for green belts, or $760 million for the total area involved.

In sum, the area of critical moving sand dunes is estimated at 4.1 million hectares. The total cost of sand dune fixation would be of the order of $900 million. No country-by-country breakdowns are available. In the absence of detailed estimates, it is assumed for the purposes of the financial plan that one-half are in developing countries requiring external financial assistance. The only countries for which estimates are available are Australia, the United States and the USSR, where the areas of moving sands are estimated at 1.6 million hectares. In view of the relatively small magnitudes involved, the margin of error would not significantly affect the total estimated costs of corrective measures.

Other essential measures

The problems of desertification are complex and require a variety of measures for their solution. This is especially true of rangelands and, to a lesser extent, of rainfed croplands, where population and animal pressure is severe. In many cases, expenditures for the proper technical management of rangelands will not succeed unless this pressure is permanently relieved. Because they are fragile ecosystems, the world's arid and semi-arid rangelands have a limited bio-mass and limited carrying capacity. Two major factors in the desertification of rangelands are overgrazing and the excessive cutting of fuel-wood. It is estimated that some 4 million hectares of open woodland (steppe and savannah) are cut every year. Reafforestation and revegetation of denuded land is an integral part of action to protect against desertification (e.g. shelter-belts), as are stabilization of productive rainfed agricultural and pastoral systems (the concept of green belts), and provision of alternative sources of income.

As was noted earlier, the financial plan presented here includes estimates of the costs of only part of the measures called for in the PACD. Carrying out the other measures that are essential to the success of the Plan would entail substantial additional costs that cannot now be estimated. It is not necessary to review all of these measures here, but it may be useful to call attention to some of them that may prove to be especially significant financially.

Alternatives to pastoral livelihood systems

Overgrazing and animal pressure can be reduced by stratification of livestock, i.e. by careful planning of the movements of livestock at various stages of their growth cycle to achieve the optimum distribution of herds throughout the different rainfall zones. This approach would involve serious political problems, especially when movements across national borders are required. Moreover, most of the world's arid and semi-arid rangelands – over 75 per cent – do not have sufficient carrying capacity under present conditions to support their present livestock populations. In many areas, livestock populations will have to be reduced and this will encounter formidable political and social obstacles. It is suggested that in order to make this transition less difficult and as an essential part of the measures to improve the condition of rangelands the following measures be taken:

(a) Intensive production of fodder. The need for animal feeds can be met more fully if fodder could be grown intensively on the perimeters of irigated areas. This would supplement the traditional grazing methods of the pastoralists.
(b) Storing animal feeds for periods of drought. Attempts should be made to build up stocks from local surpluses. Animal feeds could also be provided by developed countries as emergency assistance to poor countries such as those of the Sudano-Sahelian region. This measure should be undertaken as insurance against risk of drought and should not be used as an artificial means of maintaining herds at a level clearly beyond the carrying capacity of the land.
(c) To reduce the number of livestock in rangelands to a safe level and yet provided the pastoralists with sufficient income, it may be necessary for a time to subsidize these industries. A more viable and permanent solution, however, should be found. This would require the generation of new livelihood systems to supplement the incomes of pastoralists. Such alternative livelihood systems could include the growing of promising drought-resistant cash crops, such as gum arabic, guayule and jojoba. In other cases, where agricultural opportunities are clearly limited, the generation of livelihood systems outside the agricultural sector should be carried out. These all call for programmes of development.

Meeting the energy crisis

The cutting of firewood, while a cause of desertification, is necessary to supply the energy requirements of the people, not only in rangelands and farms, but also in urban areas. One way of meeting energy needs would be the establishment of fuel-wood plantations, especially around towns and cities in developing countries. Alternative technologies for energy may also be developed, such as bagasse charcoal and small inexpensive stoves including, in certain cases, solar stoves.

Monitoring

Another important measure is the monitoring of desertification and related natural resources processes. This would give more accurate data on the extent, location and kind of desertification taking place and its effect on the populations involved. Such information is essential for planning anti-desertification programmes and monitoring their effectiveness. With the spectacular advances made in remote sensing, regional and global monitoring systems using Landsat imagery, aerial photography and ground checking, national, regional and global monitoring systems could be established as part of the anti-desertification programme.

Total cost of financial plan

Counting only the measures whose costs can be estimated, the total cost of a 20-year programme to combat desertification in developing countries requiring external financial assistance would be on the order of $48 billion, in 1980 dollars. This figure can be broken down as follows:

(a) The rehabilitation of all irrigated lands in such countries affected by desertification, 16.3 million hectares. The total estimated costs would be on the order of $8.2 billion (low), $12.3 billion (medium) and $19.6 billion (high).
(b) The rehabilitation of 50 per cent, (or 723 million hecares) of the 1.4 billion hectares of all rangelands affected by desertification in such countries. The total estimated costs would be $7.2 billion (low), $18.1 billion (medium) and $28.9 billion (high).
(c) The rehabilitation of 70 per cent (or 68 million hectares), of the 97 million hectares of all desertified rainfed cropland in such countries. The total estimated costs would be $13.6 billion (low), $17 billion (medium), and $30.6 billion (high).
(d) Sand dune fixation in critical areas totalling two million hectares. The cost would be of the order of $449 million.

On an annual basis, the total costs of these measures, assuming that they are implemented over a period of 20 years, would be of the order of $1.5 billion (low), $2.4 billion (medium) and $4 billion (high). These are estimated average annual expenditures, and it is assumed that expenditures will be lower in the initial years and increases with time.

An alternative approach

The financial plan presented here deals with aspects of a programme to counter the degradation and declining productivity of land affected by desertification. An alternative approach is to limit the anti-desertification effort to those lands that deteriorate each year to a level of zero or negative net economic returns. The difference between these two approaches is essentially that of dealing with a process on a broad scale and defending a line.

Because it involves a smaller land area – 9.8 million hectares annually in developing countries requiring external financial assistance – the second approach would be less expensive, although the unit costs would be higher since the areas involved would be severely desertified. The estimated annual cost of comparable aspects of this approach would be of the order of $1.3 billion. It would not, however, help to stop desertification on a comprehensive basis. The degradation of the productive capability of land that had not yet fallen would continue, with adverse ecological and socio-economic consequences. Moreover, by failing to deal broadly with the desertification problem, this alternative approach would not keep additional land from falling to or below the level of zero net economic returns every year for an indefinite future.

The need for external assistance

The General Assembly has requested that, in addition to giving the components and estimated costs of a financial plan to stop further desertification, the plan include estimates of what is already being financed and what additional resources may be needed to meet the minimum objectives of stopping the spread of desertification. Estimates of the level of expenditures for combating desertification are not readily available. The reason is that anti-desertification activities are not a distinct, separately identified category in statistical classifications of expenditure flows. This is true not only of bilateral and multilateral programmes but also of national development plans and budgets, which classify their planned, on-going and past expenditures according to systems which do not identify desertification as such.

In order to obtain an estimate of the amount of external assistance concerned with desertification control to arid and semi-arid developing countries, the Desertification Unit of UNEP held discussions with officials of OECD and multilateral assistance agencies. Using the OECD project classifications, an average desertification control component for each type of project was estimated. On this basis, estimates were made of the flows of assistance from OECD Member States; international and regional financing agencies including the World Bank, the regional development banks and IFAD; bilateral and multilateral assistance from OPEC Member States; and assistance from the United Nations system other than the World Bank and IFAD.[22] The total external assistance involving desertification control has been calculated at $527 million in 1978 divided as follows:

OECD Member States plus international and regional financing agencies:	$443 million
OPEC bilateral and multilateral flows:	$23 million
UN system other than World Bank and IFAD:	$61 million

There is reason to believe that these estimates are high as they include some expenditures for areas which are not presently threatened by desertification. Taking the 1978 estimate, external assistance constitutes 22 per cent of the average annual amount of funds necessary for a minimum programme of those corrective measures whose costs can be estimated.

In developed countries, national expenditures to control desertification are significant and will probably increase. In less-developed countries, however, they are inadequate and, in the case of the least developed of the developing countries, they are negligible. On the basis of samples of on-going desertification control programmes, it is estimated tentatively that expenditures by governments of developing countries on such programmes account for roughly 10 per cent of total expenditures, including external assistance. The on-going expenditures for combating desertification in 1978 would therefore be of the order of $59 million. This represents only 2 per cent of the cost of a minimum 20-year programme.

In short, the levels of additional assistance required to stop further desertification, after taking on-going expenditures into account and ignoring measures whose costs cannot be estimated, would amount to $1.8 billion per annum, as illustrated in the following table:

Total annual requirements (medium estimate):		$2,358 million
Current level of external assistance:	$527 million	
Current level of local expenditure:	$59 million	
Total annual expenditure:		$586 million
Net additional resources required:		$1,772 million

The figures indicate that a considerable amount of additional resources are required for combating desertification. It should be emphasized that these estimates do not take into account the costs of many measures which are essential for the implementation of desertification control programmes. Nor do they include rising costs due to inflation and the increased costs which would result from the deterioration of what are now slightly or moderately desertified lands to severely degraded conditions, which would be more expensive to correct. While initial expenditures should necessarily be low and built up with time, there is a clear and immediate need to increase substantially the level of assistance to developing countries suffering from desertification.

IV METHODS FOR THE MOBILIZATION OF DOMESTIC RESOURCES

In considering the subject of the methods of mobilizing domestic resources for implementing the PACD, a distinction should be made between the developed and the developing countries. In developed countries suffering from desertification such as Australia, the United States and the USSR, the resources are available, and methods of mobilizing them have been developed and applied. It is for this reason also that the financial estimates given in Chapter II focused on the needs of the developing countries.

It is in the poorer countries, especially the least developed, that the problem of mobilizing domestic resources is particularly acute. In these countries, furthermore, the ways of mobilizing domestic resources are closely intertwined with problems of low levels of overall domestic resources available and the competing demands on the allocation of these resources.

Re-allocation of limited available resources in favour of anti-desertification projects is a delicate policy-planning matter. In those countries most seriously affected by desertification, special priority should clearly be given to desertification control projects. Priority should also be given in countries where the results of desertification are not so apparent, but where protective and corrective measures should be taken lest the deterioration accelerates and results in severe losses in future production.

It should be realized that, although desertification control projects do not necessarily generate early returns in the form of financial flows, their benefits are unquestionable. These include increased food production, the halting of declining yields due to land deterioration and the prevention of abject poverty and the loss of livelihood systems.

There are, furthermore, social and ecological benefits, including the protection and enhancement of a vital, life-sustaining part of the environment, which may not be readily quantifiable but which are essential to the well-being of the country as a whole. It is necessary, for these reasons, to have well-formulated projects providing information on these important factors and to make governments more aware of the need to give priority to desertification control programmes in their development plans and allocation of resources.

Attention is drawn, in this connection, to the guidelines provided in the PACD, particularly to the recommendations that public participation be made an integral element of the prevention and combating of desertification. Account would thus be taken of the needs, wisdom and aspirations of the people. Attention should also be given to the PACD's recommendations concerning regional projects. It is emphasized that the mobilization of local potential resources is an essential prerequisite and an integral part of some programmes which require domestic financial resources as well as, in the case of many developing countries, foreign assistance.

Domestic fiscal capacity

To estimate the potential for the mobilization of domestic resources, an analysis has been made of the fiscal capacity of the affected countries. For working purposes, the countries for which the estimates were obtained were taken from a list of the countries of the world with arid and semi-arid areas. These are shown in Table 5 (page 77). The

Table 5 List of Countries with Arid and Semi-arid Areas, by Region and Degree of Development

Region	Countries most affected by aridity and least developed	Countries most affected by aridity but not least developed	Countries less affected by aridity but least developed	Countries less affected by aridity but not least developed
Mediterranean	Democratic Republic of Yemen Yemen Arab Republic	Algeria Bahrain Egypt Iraq Israel Jordan Kuwait Lebanon Libyan Arab Jamahiriya Morocco Oman Qatar Saudi Arabia Syrian Arab Republic Tunisia Turkey United Arab Emirates		Cyprus Greece Malta Portugal Spain
Americas		Argentina Chile Mexico Peru United States of America		Bolivia Brazil Canada Colombia Ecuador Paraguay Venezuela

Table 5 *continued*

Region	Countries most affected by aridity and least developed	Countries most affected by aridity but not least developed	Countries less affected by aridity but least developed	Countries less affected by aridity but not least developed
Sudano-Sahelian and Tropical Africa	Angola Central African Republic Chad Ethiopia Mali Niger Somalia Sudan Upper Volta	French Territory of Afars and Issas (now Djibouti) Mauritania Namibia Senegal Western Sahara	Bostwana Gambia Lesotho Uganda United Republic of Tanzania	Cape Verde Ghana Kenya Madagascar Nigeria Swaziland United Republic of Cameroon Zambia
Asia and the Pacific	Afghanistan	Australia China India Iran Mongolia Pakistan USSR		Burma Sri Lanka

Source: This list was prepared by a panel of senior consultants to the Secretary-General of the United Nations Conference on Desertification (UNCOD), an *Ad Hoc* Inter-Agency Task on countries with arid and semi-arid areas and the UNCOD Secretariat.

methodology involved a survey of the fiscal indicators in these countries. The survey showed a wide range of tax revenues as a percentage of Gross Domestic Product (GDP). A number of countries demonstrated a high capacity, on the order of magnitude of 18 to 20 per cent; others showed very low capacities, on the order of magnitude of only 11 to 12 per cent. For those countries suffering from desertification with lower capacities, an attempt has been made to estimate the increased revenues that might be realized if the revenue/GDP ratio were raised. The procedure used was to consider the countries by income groups (low, medium, high) and for all countries with low capacity, a ratio equal to the average of the per capita income group was then applied. The resulting estimates are defined as the potential government revenues.

The potential government revenues, reflecting the increase that could be achieved by more efficient tax efforts, are shown in column 5 of Table 6 (page 79). As can be seen, the increase is not large. The global total of $1,550 million for the 'total most affected' countries represents an increase of only 3.2 per cent of actual revenues. (This compares with a potential increase of $2,028 million, or over 8 per cent, for the 'less affected' countries.) It is obvious that even if potential government revenues were realized, the amounts available for desertification control programmes would be limited, especially in relation to the estimated financial needs described in Chapter II.

Another method of measuring the potential level of resources which could be mobilized domestically is to measure the growth increments from which these resources could be mobilized. The aggregate real growth rate of the affected countries, as listed in Table 1, is estimated at less than 4 per cent annually, compared with 5 per cent for all developing countries,[23] so that increased incomes that can be tapped would not appear to be a significant overall source of capital mobilization. In examining the fiscal capacities and growth increments of the affected countries one must consider, furthermore, the competing claims for a

Table 6 *Basic Fiscal Capacity Indicators for Developing Countries Affected by Desertification*[a] (1977 estimates)

Breakdown by Region and Degree of Aridity	Population (millions)	Aggregate GDP (millions $US)	Actual Government Revenue (millions $US)	Actual Government Revenue as a per cent of GDP	Potential Government Revenue (millions $US)[c]
	(1)	*(2)*	*(3)*	*(4)*	*(5)*
Most Affected					
South America	27.02	24,960	6,131	24.6	6,308
Northern Africa	63.05	29,030	7,012	24.2	7,112
Sudano-Sahelian and southern neighbouring countries	86.59	18,110	2,736	15.1	3,236
Middle East	63.18	62,090	14,474	23.3	14,882
Asia	717.76	118,400	18,686	15.8	19.051
Total Most Affected	957.60	252,590	49,039	19.4	50,589
Total Less Affected[b]	234.83	99,774	25,029	25.1	27,057
Total All Affected	1,192.43	352,364	74,068	21.0	77,646

Source: World Bank, *World Tables* 1976, and IMF *International Financial Statistics.*
[a] Countries affected as given in list in Table 5. Excludes: Peoples Republic of China and Mongolia; Algeria and Mexico because of oil surplus; and Western Sahara for lack of data.
[b] Excludes Brazil which otherwise dominates the group.
[c] Computed by applying to each country the average revenue ratio in 1973 for its income group, as given by the *World Tables*, p. 440.

share of these increased revenues. Bearing in mind that domestic resources available in most developing countries are extremely limited, we note that the following are some of the principal methods by which domestic resources could be mobilized for programmes to combat desertification.

Conventional methods

A basic approach to mobilizing domestic resources is to increase the tax efforts. In the area of administration the system of tax collection can be improved by strengthening the tax authorities of the government. More efficient revenue collection could be realized in this connection, by providing local governments with taxing powers.

Income taxes in developing countries usually account for a small portion of revenues because the potential is quite low. The existence of high income groups does sometimes permit a far greater tax intake than average income might suggest, but this is not the case in many desert-prone developing countries, where such groups are few or small. For lower income groups in many of the affected countries, especially in the Sudano-Sahelian zone and southern neighbouring countries, income taxes are non-existent or non-applicable, and poll taxes (or hearth taxes, or the *impôt minimum fiscal*) are applied instead. Their volume, however, is generally small and has been either considerably reduced or largely eliminated since the droughts of the early 1970s decimated the potential to pay.

Other possible taxes are indirect taxes (sales or turnover, value-added, excise, customs duties on both imports and exports). These, however, already account for the major portion of revenues (60 to 70 per cent), and increases may cause serious disincentives and other counter-productive effects.

Betterment taxes at first sight would seem to be very appropriate to the type of projects involved in combating desertification, for these projects generally have high social returns. Taxing such an 'externality benefit' – in this case the increase in private land values due to social, publicly-financed projects – is the basic rationale of betterment levies. In general, however, such taxes are difficult to administer in rural areas and are more effective in urban communities. Capital improvements resulting from increased production through anti-desertification measures, furthermore, are usually realized over the long-term. Much grazing land, especially in the Sahel, is considered part of the public domain, and is therefore not taxable.

The most promising source of tax revenues appears to lie in the area of user charges, such as in irrigated areas, cattle ranching schemes, rangeland wells, veterinary activities and woodlots. All of these show tangible benefits readily identifiable to the users. It should be realized, however, that such taxes can be made only after initial investments have produced the necessary result. Even in this area, moreover, there are problems, to be encountered such as initial resistance to 'purchases' which may have originally been 'freely available', problems of administration and poorly functioning facilities.

In sum, there is clearly a potential for increasing domestic revenues to finance desertification control programmes, but this potential appears to be limited by the limited fiscal capacities of countries, limited domestic resources, difficulties in administration and sociological problems. The most promising tax measures for raising revenues appear to be in the field of user charges.

Promotion of viable commercial schemes

Another possible method of generating domestic sources of income lies in the promotion and development of commercially-viable enterprises with initial technical or investment assistance. These could include such schemes as feed-lots and fattening ranches, bio-gas production at feed-lots, forage production, desert tourism and – probably the most promising – the development of certain cash crops.

It may be noted that the guayule and the jojoba, which are native to the arid and semi-arid areas of southwestern United States and northern Mexico, and gum-arabic (*Acacia senegalensis*), which grows in many parts of Africa, thrive in arid and semi-arid lands and therefore could be grown extensively in the desert-prone areas of the world. There have been successful experiments in growing and transplanting these crops. In addition to their suitability to the semi-desert ecosystems, these crops have significant potential and export markets. The guayule is a source of latex and rubber; jojoba produces a valuable kind of oil found elsewhere only in the sperm whale, an endangered species; and gum-arabic has a variety of uses in the pharmaceutical, confectionery and other industries. These crops could thus serve the dual purpoe of combating desertification and generating cash revenues. Commerically profitable enterprises such as these could be used for mobilizing domestic capital either through taxation or through the ploughing back of profits for investment.

Still another method is the use of a 'food for workers' scheme for the mobilization of a country's labour resources. This, of course, would depend largely on contributions by food surplus countries of the necessary food supplies. If food (from external sources) were provided during the off-season when it would be more valuable to workers and thus represent higher wages in kind, labour resources could be more readily mobilized. There are already many valuable examples of such projects, in particular in the Sudano-Sahelian zone.

Injection of external resources

It is not realistic to expect the poorer countries affected by desertification to mobilize more than a small fraction of the domestic resources required for desertification control through their own efforts. A large part of these resources must be mobilized through the injection of foreign aid.

As has been done in many past aid programmes, aid donors can provide goods marketable in the recipient country. The local currency obtained by selling such goods can then be put in a special account and used to finance anti-desertification projects. A variant of this approach would be for aid donors to accept local currency, rather than convertible foreign exchange, for part of their exports to developing countries affected by desertification. This local currency could then be put in a special account and used for anti-desertification purposes.

External resources can also be distributed on a preferential basis to provide incentives to relocate populations and economic activities (such as charcoal-making) in ways that facilitate the control of desertification.

National development financing institutions

In general, national governments do not make loans directly to development projects, such as those concerned with desertification control. Loans are usually made through semi-autonomous or quasi-governmental bodies, such as national development banks, marketing boards, development corporations and other institutions in which the governments have an equity. In the poorer countries suffering from desertification there is a dearth of such institutions, and the financial resources of existing institutions are limited.

Many development banking institutions – despite their mandate to promote development projects – would not be apt to finance desertification control projects, except those which are likely to yield early cash returns which could be ploughed back into the capital resources of these banks. In many countries suffering from desertification, furthermore, especially in the Sudano-Sahelian zone and southern neighbouring countries, there are few development banks and those that exist have little capital.

When considering loan applications, the national lending institutions appraise the project proposals on the basis of a number of criteria. One of the major ones is the question of priority, in view of the competing demands for the scarce resources. Unfortunately, desertification control is not given as high priority as it should be, except perhaps in a few of the most seriously-affected countries. The reason, partly, is that control projects often do not yield returns in the form of early cash flow, which other projects do. Desertification control projects are also often neglected because of their remoteness from the centres of decision-making. Desertification, furthermore, has not to date been categorized as a specific sector in economic and related classifications of government and international institutions.

In view of the above, desertification control projects in general do not receive loans from national lending institutions on a priority basis. The situation, however, can be improved in various ways. Although desertification is not considered a sector as such, actual anti-desertification projects are concerned with such goals as agricultural improvement, rural development, food production, development of water supplies, development of cash crops and other designations with fall within the priorities of development planners and financial decision-makers. These aspects of the projects should therefore be highlighted together with the social and ecological benefits noted at the beginning of this chapter.

V THE PRACTICABILITY OF OBTAINING LOANS FROM GOVERNMENTS AND WORLD CAPITAL MARKETS ON A CONCESSIONARY BASIS

Loans from foreign government sources

As a general rule, loans on a concessionary basis are provided to the poorest countries, which are least able to borrow on commercial terms. These are mainly the least developed of the developing countries (LDC), which have been classified as such by the United Nations General Assembly because of their low per capita incomes, limited natural resources, lack of economic diversification or other factors. Other bases for the provision of loans on a concessionary basis are the degrees to which the countries have been 'most severely affected' by the energy crisis. In addition, certain bilateral assistance agencies accord priority to certain regions because of the severity of their geographical problems (e.g. the Sudano-Sahelian region), or because of special ties (e.g. the assistance programmes of Australia and New Zealand in the South Pacific and the French assistance programmes in various parts of Africa).

An important factor in obtaining loans from government sources is the cost-benefit relation in the particular desertification control projects.[24] Most governmental and intergovernmental financial institutions (e.g. IDA) apply technical feasibility and cost-benefit analyses in their appraisals of project proposals, even when loans are provided on a concessionary basis, first, to ensure that the project is technically feasible and, secondly, to ensure that it will have a positive impact on development. Increasing recognition is now being given to social costs and benefits and to the concept of national profitability, rather than strictly to the financial profitability of the project itself. There is also increasing attention given to the benefits of the projects to the poorer segments of the populations of the countries, especially in the rural areas.

Loans from foreign governments are generally made through statutory bilateral assistance agencies or development corporations. Most bilateral assistance programmes in the international donor community provide grants and loans on highly concessionary terms as part of their official development assistance. A number of bilateral government assistance programmes, although they do not include desertification as a separate sectoral classification, do give priority to desert-prone arid and semi-arid areas, such as the Sahel. Since desertification control programmes, furthermore, involve rural development and affect the poorest segments of the population, they would be looked upon favourably by foreign lending agencies for lending on concessionary terms.

It is important, nevertheless, to have well-prepared applications which describe the favourable benefit-cost ratios and social benefits of projects from a national viewpoint, as well as the high priority that the recipient government accords them, reflected, to the extent possible, by pledges of the necessary counterpart support. In applying for assistance, the following points should be highlighted:

(a) Desertification-control projects support priority objectives of bilateral assistance programmes, such as agricultural improvement rural development, food production, water resources development and basic needs.
(b) Rates of return, even if not in the form of early financial profits, are often high. Studies undertaken for the UNCOD (1977) showed internal rates of return ranging from 14 per cent to 51 per cent for projects financed by the World Bank and the Inter-American Development Bank. Broad estimates also indicated potential benefit-cost ratios in order of magnitude of over 3.[25]
(c) Desertification control projects are typically in rural areas and benefit the poorest segments of the population.

While desertification control projects, for the reasons explained above, would be looked upon favourably for lending on a concessionary basis, it should be noted that the current international financial environment indicates a slowing down or even a reduction of the amount of development assistance provided by the donor community, particularly the OECD countries. While the OECD countries as a group ran substantial current account surpluses in the 1960s and early 1970s, the situation radically changed in 1974, which was followed by years of large deficits, although there was a small surplus in 1978. In 1979, a current account deficit of about US$30 billion was recorded by the OECD countries, and present estimates suggest a current account deficit of US$74 billion for 1980 (the exceptions are expected to be Switzerland and Norway). The OPEC countries, on the other hand have had a parallel and somewhat larger surplus. The 1975–78 trend of declining OPEC surpluses was reversed in 1979, when it totalled US$60 billion. Preliminary estimates indicate that the OPEC surplus will be about US$115 billion in 1980 [the 1985–6 drop in oil prices changed this situation and hence affected the aid capacity of OPEC countries], a level which is not expected to decline rapidly in the succeeding years.[26] It may be noted that the official development assistance of OPEC countries amounted to $5.3 billion, or 1.59 per cent of their GNP, in 1978.

Although there are already many competing demands for both OECD and OPEC funds, most of the OPEC countries are themselves suffering from desertification and OECD countries have an interest in providing assistance to less developed countries, including those suffering from desertification. Both groups, therefore, might be interested in participating and taking the lead in international programmes for combating the degradation of and for the development of arid and semi-arid lands.

Loans from world capital markets

By definition, there is an inherent paradox in obtaining loans from world capital markets on a concessionary basis. The reason is that the world's capital markets are commercial and profit-making in nature, so that they do not provide loans on concessionary terms. The principal means, therefore, by which concessionary terms could be obtained for the borrower would be through assistance from third parties through mechanisms such as guarantees of interest and principal, payment of part of the interest charges, advancing the amortization of the loan and blending concessionary and non-concessionary funds. In this way, the borrower could obtain concessionary terms.

In principle, grants by aid donors could make it possible to re-lend on concessionary terms funds borrowed on capital markets. If such grants are insufficient, some of the new means of raising funds discussed in Chapter I might be considered. For example, if the international community decided that it was appropriate, a tax might be levied on non-renewable energy resources. Whatever source of revenue is used, some entity would have to be charged with the task of allocating the funds raised.

The most prevalent means of borrowing from world capital markets are through investment banking, or underwriting, and through commercial bank loans. Investment banks raise the funds both through public offerings and private placements. It should be emphasized that they are more interested in the nature of the guarantees of repayment, the assurances of continuity and reliability, i.e. automaticity, and other standards of creditworthiness, than in the nature of particular projects. Commercial banks, unlike investment banks, are primarily interested in lending to governments or other borrowers for individual projects directly rather than through an intermediary institution for a group of projects. Their credit standards, however, are similarly rigorous.

To obtain loans on concessionary bases from world capital markets, it is unlikely that a developing country would negotiate such a deal directly with a commercial bank or other private source of financing, and the use of an intermediary development finance institution would be indicated. There is a limited number of such institutions, e.g. the World Bank, IDA and the regional development banks, which could 'package' concessionary loans for the developing countries, the sources of private capital and the donor willing to blend grant and long-term, low-interest development funds. Such deals could also be arranged by a public international financing corporation, such as was suggested by the study contained in Part I. The next chapter of this study will deal with the proposal for the establishment of such an institution, and Section V of Part III presents a feasibility study on its establishment.

VI A PROPOSED PUBLIC INTERNATIONAL CORPORATION FOR FINANCING ANTI-DESERTIFICATION PROJECTS

Background and nature of proposal

The establishment of a public international corporation to attract investments and provide financing for suitable anti-desertification projects with non-commercial rates of return was proposed by the UNEP expert group on additional measures (Part I). The proposal called for providing the corporation with equity funds from countries with surpluses in their international accounts, as well as from financial institutions.

The original proposal is understood as relating essentially to those projects which, although indispensable to the very survival of human and animal life in the arid and semi-arid regions, have extremely long gestation periods and, while having substantial expected benefits, present unusual difficulty in quantifying the potential returns at the preparation stage of the project. A large number of desertification control projects will be of this character. A coherent financial plan requires that provision be made for their financing.

Feasibility

The feasibility of establishing such a public international corporation depends on whether donor countries and organizations will provide the necessary resources for its establishment.

The projects which the corporation will be expected to finance would for the most part be incapable of bearing interest costs, even on the highly concessional terms which are presently available through IDA and similar financing institutions. It is important to recognize, therefore, that such projects will have to be financed primarily with funds provided on an interest-free basis.

First, the financial institutions such as IDA, IDB, IFAD and the regional development banks should be encouraged to expand their activities to financing anti-desertification projects. It is recognized, however, that there will still remain a substantial number of projects which these institutions could not reasonably be expected to finance, and hence there is room for an institution whose main function will be the financing of those long-term projects of desertification control with non-commercial rates of return, which would not ordinarily commend themselves to existing financial institutions.

If the proposed new financing institution were established in the form of a public international corporation, it would have the flexibility in operation, and would encourage the efficiency in management, that would commend its activities to potential suppliers of funds.

Assuming efficient managerial leadership, a public international corporation established for the primary purpose referred to above should readily attract a nucleus of uniquely-qualified and competent staff with a degree of specialized expertise in the anti-desertification field not possessed by other financial intitutions. It should not, therefore, be precluded from extending the range of its services to include projects which are capable of being funded on a basis which will assure over time a return of capital and even, in some cases, modest interest charges. To the extent that the corporation were able to establish a sound record of managing such projects, it would be reasonable to expect that it would attract funds from sources which would not be interested if the activities of the corporation were confined exclusively to undertakings which could be financed only on a full grant-in-aid or interest-free loan basis.

The Group is fully aware of the reluctance with which the Member States of the United Nations receive proposals for the creation of new institutions. It suggests, therefore, that if the United Nations decides that funds should be provided for the establishment of the proposed public international corporation, the possibility of having this corporation made a subsidiary or affiliate of an existing institution should be actively explored.

Establishment as an affiliate or subsidiary of an existing institution

The establishment of such a corporation as an affiliate would have the advantage of lending it the reputation, facilities and expertise of the established institution. In this way, it would also benefit from a shortening of the proving period. By a decision of the competent organ of the intergovernmental financing organization, combined possibly with an agreement with donor governments and institutions, the proposed corporation could be established. A corporation which is established as an affiliate of an existing institution could also have the advantage of eligibility (in determined circumstances) for the supply of capital from the parent institution as well as from donor members.

There are a number of possibilities for such an affiliation role. One possibility is the World Bank, with its solid reputation. It may be noted that the Bank is already engaged in a variety of activities and has major affiliates – IDA, for which it raises replenishment funds and contributes a portion of its operating income, as well as the International Finance Corporation (IFC).

Another possibility which could be pursued is the OPEC Fund. Many of the major OPEC members have serious problems of desertification and, therefore, should be particularly interested in the subject. Their interest could be heightened if the purposes of the public international corporation were to include not only combating desertification *per se* and reclaiming desertified land, but also desert use and development – all of which are intertwined. There is a movement to transform the OPEC Fund into a development agency capitalized at $20 billion, with authority to borrow in capital markets to re-lend for a variety of development purposes, including the financing of balance-of-payment deficits, project financing and certain merchant banking functions. The Fund will have currently at its disposal a cash amount of $2.4 billion after the current round of replenishment is completed, of which around $2 billion is likely to remain undisbursed.

A third possibility is IFAD, which started operations in 1977 with a subscribed capital of $1 billion. Although a new organization, IFAD is progressively gaining experience in its loan operations and has given priority to assisting poor countries, especially the poorer segments of their rural areas. Desertification control, therefore, could be of possible interest to IFAD.

Establishment of an independent institution

If none of the organizations cited above would be willing to establish an affiliate organization and it were not possible to find another appropriate capital financing agency for the establishment of a public international corporation with affiliate status, the proposed corporation could be established as an independent institution. The feasibility of its establishment, as noted previously, will depend on the willingness of the international community to provide the necessary capital.

Independent or affiliate

A donor government or institution wishing to make a contribution to the resources of the corporation in the form of funds to be applied towards the subsidization of interest rather than as capital, could also do so. The operating and administrative costs of the corporation could be provided from its paid in capital, and from interest and operating income supplemented initially by grants from governments or organizations such as the UNDP, which has provided considerable institutional support in meeting the administrative costs of the Caribbean Development Bank.

The proposed corporation could either borrow directly by issuing notes or bonds through investment banking firms, or expedite borrowing by the governments of affected countries from commercial banks. Notes or bonds issued by the corporation would be guaranteed by the governments investing capital in it. Funds raised at commercial interest rates could then be blended with concessionary loans and grants to lower the average rate for each project. This would be accomplished by package arrangements including 'hard' and 'soft' loans combined with grant elements.

Summary

The possibility of encouraging existing institutions to finance anti-desertification projects should be actively explored. As this approach might not appear to be ample, the creation of a public international corporation to channel funds to anti-desertification projects deserves serious consideration. Such a corporation should, if possible, be established as an affiliate of an existing financial institution.

VII INCREASING THE ROLE OF FOUNDATIONS IN ANTI-DESERTIFICATION RESEARCH AND TRAINING

General Assembly resolution 34/184 requested that the study should deal, *inter alia*, with 'the means for encouraging the active participation of foundations in the financing of anti-desertification training and research programmes'. It is assumed that the Assembly had primarily in mind the institutions known as private foundations, of which the majority are to be found in the developed OECD countries – the United States, Canada, the countries of Western Europe, Japan, Australia and New Zealand.

In the United States, foundations, of which there are over 26,000, are classified as (a) independent; (b) company-sponsored; (c) operating; or (d) community.[27] A few government foundations, such as the National Science Foundation, also make grants for specific purposes. In terms of their resources, the private foundations have assets in the region of $32 billion; in 1978, they made grants of over $2.1 billion. In Europe the number of private foundations is smaller, and so is the average size of the assets they control. Between 16,000 and 21,000 foundations possessing assets valued at $8.5 billion made grants of about $900 million in 1978. In other OECD countries about 1,000 foundations had assets of some $300 million.[28]

A new kind of agency dispensing assistance for training and research has recently emerged. Some are government agencies, such as Canada's International Development Research Center (IDRC), the Swedish Agency for Research Co-operation with Developing Countries (SAREC), and the United States' Appropriate Technology International and Inter-American Foundation. These are oriented towards indigenous research and training projects in developing countries. Their annual grant-making resources are of the order of $100 to $150 million.

In addition, there are the four 'political' foundations in the Federal Republic of Germany, representing the interests of four major political parties and receiving government funds in proportion to the size of their representation in the national legislature. In the aggregate, these foundations have an annual budget of $130 million, of which about $50 million are allocated for international purposes.

In a somewhat different category are a number of foundations recently established in the Middle East as non-governmental bodies serving to express the philanthropic interests of their founders. These include the Queen Alia Foundation (Amman), the King Faisal Philanthropic Foundation (Riyadh), and the Philanthropic Trust (Kuwait). There are no available estimates of their resources, but these foundations could possibly be interested in financing research and training related to desertification control.

Fields of interest

A survey of United States' foundations in 1978 showed the following distribution of the grants they made by dollar volume:[29]

Field	Percentage of Total
Education	27
Medicine/Health	21
Social and Natural Sciences	17
Social Welfare	15
Arts and Humanities	10
International Activities	8
Religion	2

In Europe, the distribution, measured by numbers of foundations involved in each area, was as follows:[30]

Field	Percentage of Total
Medicine	19
Social and Natural Sciences	17
Arts and Humanities	14
Social Welfare	12
International Activities	9
Education	6
More than one field	23

It will be seen, therefore, that anti-desertification research and training is by no means a priority interest of the foundations at this time.

The foundations which are most likely to be associated with the anti-desertification effort are those which already have an interest in one or the other of the following two fields: international relations and the environment. The possibility exists, however, that a few foundations whose interests lie in the fields of education and the earth-sciences may be encouraged to participate.

Private voluntary agencies in the OECD countries, have, over the years, made substantial assistance grants to developing countries. Table 7 (page 92), which is extracted from the OECD review of 1979, shows

Table 7 *Development Assistance Grants by Operating Foundations in the OECD Countries in 1978*

	($ millions)
Australia	38.3
Austria	14.6
Belgium	31.5
Canada	87.0
Denmark	8.1
Finland	6.2
France	19.9
Germany	284.0
Italy	0.3
Japan	18.9
Netherlands	55.5
New Zealand	5.9
Norway	25.9
Sweden	44.3
Switzerland	48.6
United Kingdom	50.2
United States of America	924.0
	1,663.2

Source: OECD, *1979 Review: Development Co-operation* (OECD, Paris, 1979).

that the amounts of such grants equalled $1,663 million in 1978. It can, therefore, be assumed that such philanthropic institutions can be persuaded to take an interest in anti-desertification research and training, especially since such activities as agricultural development, arid and semi-arid land development, food production, rural development, research on new plant species and others already fall within the ambit of their activity.

Potential resources for training and research related to desertification control

As far as the privately-endowed, independent foundations are concerned, no more than 150 to 175 are active international givers, and fewer than 20 account for 90 per cent of the grants to developing countries. Whilst the range of the individual grants varies from $5,000 to $1 million, the average size is considerable, between $40,000 and $300,000.

Among the better known international foundations with an interest in the Third World are the following: Ford, Rockefeller, Rockefeller

Brothers Fund, Lilly Endowment, W. K. Kellogg, Carnegie Corporation, A. W. Mellon, E. M. Clark and Tinker in the United States; Wolfson, Nuffield and Leverhulme in the United Kingdom; Alfried Krupp von Bohlen und Halbach, Kübel and Volkswagenwerk in the Federal Republic of Germany; NOVIB and NUFFIC in the Netherlands; Calouste Gulbenkian in Portugal; Juan March in Spain; and Toyota in Japan. A breakdown of some of the world's largest privately-endowed foundations, by region, assets, types of grants and areas of operation, is shown in Table 8 (page 94).

Enlisting the participation of prospective donors

As can be seen from the above, there are considerable resources available in the foundation sector which could be usefully applied to research and training programmes relating to desertification control. It should be emphasized, however, that these grant-making bodies do not ordinarily prepare their own projects. They prefer to contribute to the financing of projects brought to them by other bodies.

In order to enlist their support for programmes of anti-desertification training and research, it is suggested that the General Assembly take the following action:

(a) call upon UNEP, in the implementation of the PACD, to institute, in cooperation with the relevant United Nations Agencies, specific programmes of research and training at the national, regional and international levels;

(b) invite private foundations and other grant-making institutions in the developed countries and elsewhere to co-operate with UNEP in the execution of such training and research programmes;

(c) encourage those member governments which are not already doing so to recognize as eligible for tax exemptions private foundations which contribute to the support of international training and research programmes, specifically those relating to the control of desertification.

It is further suggested that UNEP and on its behalf UNSO, which are the entities in the United Nations system most concerned with desertification, should contact individually and systematically those foundations with an interest in international relations and environmental problems,[31] in order to familiarize them with the problems of desertification and to explore their interests in the financing and implementation of research and training.

The Consultative Group on International Agricultural Research (CGIAR), the membership of which includes foundations as well as governments and intergovernmental organizations, provides substantial

Table 8 *Largest Independent International Grant-Making Foundations (by country)*

	Assets	Total Grants	International Amount	Grants* Number	Special geographic or functional area of interest, if any
	(U.S.$ millions)				
NORTH AMERICA					
United States of America					
Ford	2,291	147	32.1	(298)	
W. K. Kellogg	827	42	13.0	(45)	Latin America
Rockefeller	740	43	11.0	(105)	
A. W. Mellon	776	41	6.1	(20)	U.S. Int'l
DeRance	103	7	5.4	(159)	(religious-welfare)
Rockefeller Bros.	182	17	3.2	(57)	Asia; Latin America
E. M. Clark	192	11	2.3	(33)	
Lilly	524	32	1.7	(40)	Africa
Luce	68	2	1.0	(11)	Asia
Tinker	29	1	1.0	(33)	Latin America
Carnegie Corp.	284	13	.8	(7)	Africa; Oceania; Caribbean (education)
Macy	47	2	.7	(31)	Medical
Kresge	586	25	.6	(9)	U.S. Int'l (Capital costs)
Subtotal	6,649	383	78.9		
EUROPE					
United Kingdom					
Wolfson	133	8	1.2	(est.)	Commonwealth
Nuffield	51	4	.6	(est.)	Commonwealth
Leverhulme	—	3	.5	(est.)	

Federal Republic of Germany			
Volkswagenwerk	763	47	3.5
Kübel	—	2	1.0 (29) (est.)
Netherlands			
NUFFIC	—	6	3.0 (est.)
NOVIB	—	5	2.5 (est.)
Portugal			
C. Gulbenkian	293 (1970)	13 (1970)	2.0 (1970 Middle East
Spain			
Juan March	65 (1971 est.)	6 (1971 est.)	.6 (1971 est.) Latin America
Switzerland	9	—	.5 Malnutrition
Subtotal	1,314	94	15.4
Total	7,963	477	94.3

* International refers to developing countries.

Sources: *The Foundation Directory*, Seventh Edition (New York, The Foundation Center, 1979); *Guide to European Foundations*, Third Edition (Turin, Giovanni Agnelli Foundation, December 1979); and *The International Foundation Directory*, Second Edition, H. V. Hodson (ed.) (London, Europa Publications Ltd, 1979).

assistance to research and training institutes which deal in whole or in part with arid and semi-arid areas and thus cover desertification control. These institutes include the International Center for Agricultural Research in the Dry Areas (ICARDA), the International Crops Research Institute for the Semi-Arid Tropics (ICRISAT), the International Livestock Center for Africa (ILCA), and the International Food Policy Research Institute (IFPRI). The CGIAR should be encouraged to increase the assistance of its members to desertification control research and training.

While recognizing the value of the activities of other United Nations agencies, it would be particularly desirable for the United Nations University (UNU) and the United Nations Institute for Training and Research (UNITAR) to increase and intensify their activities for training and research projects concerning desertification control and the development of arid and semi-arid lands. In particular, the UNU, with endowment funds of $95 million and pledges of an additional $50 million, already has a project within its natural resources programme for research and training relating to arid lands. All three of the principal areas in which UNU has decided to concentrate its efforts – the World Hunger Programme, Natural Resources, and Social Development – have relevance for the problem of desertification. UNU, with its worldwide network of affiliated institutions, is in a good position to support expanded activities in the training and research fields relating to arid lands, including desertification control.

However useful the role of foundations might become as financial supporters of training and/or research schemes in the context of national and international efforts to halt desertification, it seems clear that their contributions will not be sufficient to cover all the needs of the countries affected by desertification.

Greatly expanded research and training are fundamental to desertification control and, therefore, require major attention and greatly expanded financing, with particular emphasis on the development of national institutions in the areas subject to desertification. Additional resources will have to be devoted to the purpose: for instance, from within overall national training and research programmes, through additional government-to-government aid, or through the diversion of resources which are, at present, being used for other purposes. Some rearrangement of priorities would thus become necessary.

Notes

1 The financing system itself was terminated in the past form and shape and established within UNDP.
2 This position is reflected in the Informal Single Negotiating Text for the

Third United Nations Conference on the Law of the Sea (United Nations Document A/CONF.62/WP.8/Part I, May 1975). See also *Informal Composite Negotiating Text* Rev.2 (A/CONF.62/WP.10/REV.2, 11 April 1980).

3   Similar appeals have been made by the General Assembly in resolution 914 (X), adopted in December 1955; resolution 1837 (XVII), adopted in December 1962; resolution 2387 (XXIII), adopted in November 1968; resolution 2602 (XXIV), adopted in December 1969; resolution 2667 (XXV), adopted in December 1970; resolution 2685 (XXV), adopted in December 1970 (which called for a close link between Disarmament and the Development Decade); resolution 3470 (XXX), adopted in December 1975; resolution 31/68, adopted in December 1976; and resolution 34/88, I (h), adopted in December 1979.

4   These included a report by a group of experts appointed by the Secretary-General, 'International Compensation for Fluctuations in Commodity Trade' (E/3447, January 1961); Secretariat studies, 'Stabilization of Export Proceeds Through a Development Insurance Fund' (E/CN.13/43 of January 1962), presented to the Tenth Session of the Commission on International Commodity Trade, and 'A Development Insurance Fund for Single Commodities' (E/CN.13/45, February 1962) which was presented to the Joint Session of the United Nations Commission on International Commodity Trade and the FAO Committee on Commodity Problems; a report of a technical working group, 'Compensatory Financial Measures to Offset Fluctuations in the Export Income of Primary Producing Countries' (E/CN.13/56, January 1963), presented to the Eleventh Session of the Commission on International Commodity Trade, and an IMF report, 'Compensatory Financing of Export Fluctuations' (Washington, DC, February 1963) prepared in response to an invitation by the United Nations Commission International Commodity Trade. The report noted that the IMF regraded deficits arising out of commodity export shortfalls as a legitimate reason for the use of IMF resources and expressed its readiness to assist in the solution of problems within the context of its activities for assisting countries with balance-of-payment difficulties.

5   See Documents A/8985 and Add. 1, General Assembly, Official Records of the Fifth Committee, 1430th meeting, 13 October 1971 and A/9450.

6   Excluding Antarctica.

7   Estimates are based on UNESCO sources, including map of World Distribution of Arid Regions (1977) and Man and the Biosphere Technical Note 7 (UNESCO, 1977) as well as secretariat of the United Nations Conference on Desertification (UNCOD), *Desertification: An Overview* (A/CONF. 74/1/Rev. 1).

8   Estimate based on FAO, *Agriculture: Toward 2000* (C79/24), July 1979.

9   Estimates of irrigated land area are derived from *FAO Production Yearbook* (1978), adjusted for arid and semi-arid regions only. Other estimates are based on data compiled by H. E. Dregne, Director of the International Center for Arid and Semi-Arid Land Studies (ICASALS), Texas Technical University, Lubbock, Texas, and Consultant to FAO; and estimates

prepared by the United Nations Sudano-Sahelian Office (UNSO). Consultations were also held with Dr M. Kassas, Professor of Plant Ecology, University of Cairo. For details, see Table 1.
10 It has been calculated, on the basis of estimates provided by H. E. Dregne, that the value of the current annual loss of production due to past desertification is on the order of over $26 billion, of which $6.6 billion are from losses in irrigated lands, $7.4 billion in rangelands and $12.3 billion is rainfed croplands.
11 See *Desertification: An Overview*, op. cit., for estimates prepared for UNCOD. New estimates of land deteriorating to a level of zero or negative net returns are from H. E. Dregne, *Desertification of Arid Lands* (New York, Harwood Academic Publishers, 1982).
12 Arid lands are defined roughly as those receiving 0–25 centimetres and semi-arid as those receiving 25–50 centimetres of rainfall annually.
13 Between 1960 and 1970, Japan lost 7.3 per cent of its agricultural land to buildings and roads, European countries lost between 1.5 per cent (Norway) and 4.3 per cent (The Netherlands). See the *State of the Environment: An appraisal of Economic Conditions and Trends in OECD countries* (OECD, 1979).
14 Optimum productivity is the maximum productivity sustainable over a long period of time.
15 These estimates are based on actual costs of samples of projects supported by the World Bank, the Inter-American Development Bank and FAO. The costs of rehabilitating water-logged and salinized irrigated land through drainage and leaching varies from country to country, and project to project. For instance, recent projects, using tiles and pumps to drain water, have cost $925 per hectare in Egypt. In simpler projects using local labour for constructing open ditches, projects have been carried out at a cost as low as $120 per hectare, such as in Pakistan and the Yemen Arab Republic. A medium figure of $750 per hectare is considered to represent the majority of cases where tile drainage and leaching – but not pumping – are involved.
16 These estimates are based mainly on costs of rangeland rehabilitation projects and consultations with FAO, Work Bank and UNDP staff, as well as Balba, Dregne and Kassas. FAO cost estimates of projects involving rotational grazing and seeding are in the range of $10 to $20 per hectare, depending on the intensiveness of the practices or measures carried out. These are conservative estimates made over five years ago and do not take inflation into account. The high estimate of $40 per hectare includes control of noxious weeds by mechanical and/or chemical means. It is considered that a unit cost of $25 per hectare is a conservative average estimate reflecting the most common measures of rangeland rehabilitation.
17 H. E. Dregne, 1982, see note 11 above.
18 It is estimated that 660 million people live within the arid and semi-arid regions where the food-producing systems are threatened to a greater or lesser degree by the processes of desertification. Of this number, 62 million, living in rural areas, are immediately menaced by loss of the

productivity of the land and face the grim prospect of uprooting themselves and migrating to other areas. These are estimates of 1980; more recent assessment (UNEP, 1984) estimates that threatened people are in excess of 850 millions.

19   Such a programme of agricultural development for 90 developing countries is detailed in FAO, *Agriculture: Toward 2000*, op. cit.
20   These costings are based mainly on data on project samples from FAO. The low estimate of $200 per hectare involves minimal levels of structural changes, mainly strip cropping, residue management and contouring. Where major structural changes are involved, such as terracing, wind breaks, water conservation, contour ploughing and strip cropping, projects costs have averaged $450 per hectare. In most cases, however, less intensive techniques are adequate and the unit costs per hectare have been estimated at $250.
21   Estimates are from H. E. Dregne.
22   It may also be noted that during a series of United Nations Sudano-Sahelian Office (UNSO) planning and programming missions which visited 13 Sudano-Sahelian countries, UNSO received requests for assistance to 107 priority anti-desertification projects costing $640 million, of which $395 million had already been pledged or provided, and an additional $246 million was immediately required. The missions were undertaken in 1979.
23   World Bank, *World Development Report 1979*, Table II.
24   UNEP is conducting an extensive study of the cost-benefit analysis of environmental protection measures, which will include a study of the methodology of cost/benefit analysis as applied to anti-desertification projects.
25   UNCOD, *Economic and Financial Aspects of the Plan of Action to Combat Desertification*, A/CONF. 74/3/Add. 2, Tables II and III.
26   See International Monetary Fund, *World Economic Outlook*, especially Table 12 (Washington, DC, May 1980).
27   Operating foundations do not, as a rule, make grants but spend their money on programmes they themselves conduct. Community foundations confine their charitable activities to specific geographical areas in the United States.
28   Private foundations constitute a small portion only of the non-profit community. Church-related and secular voluntary agencies have extensive networks of overseas personnel and missions and are estimated to spend about $500 million each year.
29   *The Foundation Directory, 7th edition* (The Foundation Center, New York, 1979).
30   *Guide to European Foundations*, Third Edition (Turin, Giovanni Agnelli Foundation, December 1978).
31   *The Foundation Directory*, published biennially by the Foundation Center, lists the major United States Foundations, the value of their assets and the amounts of grants they made in the most recent year. It also indicates their main fields of interest and gives the names and addresses of the officials who should be approached when seeking grants. Approximately

similar information is given about European foundations in the *Guide to European Foundations* by the Giovanni Agnelli Foundation.

List of participants in group of high-level specialists in international financing who prepared the study contained in Part II

(*Serving in Personal Capacities*)

Rodrigo Botero (Chairman)
Former Minister of Finance of Colombia,
Former Member, Independent Commission on International Development Issues
(Brandt Commission)

Roberto Campos
Abassador Extraordinary and Plenipotentiary of Brazil to the Court of St James;
Former Minister for Planning and Coordination, Government of Brazil

George Davidson
Former Under-Secretary General for Administration and Management, United Nations (1972–1979)

Dr I. Imady
Director-General and Chairman of the Board of Directors,
Arab Fund, Kuwait

Anders Forsse
Director General,
Swedish International Development Agency (SIDA)

Thiemoko Marc Garango
Ambassador Extraordinary and Plenipotentiary of Upper Volta to the Federal Republic of Germany

A. A. Hegazy
Former Prime Minister and Minister of Finance,
Government of Egypt

Paul-Marc Henry
Former Assistant Administrator, UNEP, and President, Development Centre, OECD

Mansour Khalid
Former Foreign Minister,
Government of Sudan

Lal Jayawardena,
Ambassador Extraordinary and Plenipotentiary of Sri Lanka to Belgium, Netherlands, Luxembourg and the European Community;

Former Secretary to the Treasury, and Secretary, Ministry of Finance and Planning, Sri Lanka

Paul Gerin-Lajoie
President, Projecto International Inc., Montreal, Quebec, Canada; Former President, Canadian International Development Agency (CIDA)

Professor John W. Mellor
Director, International Food Policy Research Institute (IFPRI)

Philip Ndegwa
Chairman, Kenya Commercial Bank, Former Deputy Assistant Executive Director (Programme), United Nations Environment Programme

Keichi Oshima
Professor of Nuclear Engineering, University of Tokyo
Former Director for Science Technology and Industry (OECD)

Edgard Pisani
Senator and Former Minister of Agriculture, Government of France
Member of the European Assembly, Former Member, Independent Commission on International Development Issues (Brandt Commission)

Sir Egerton Richardson
Former Financial Secretary of Jamaica
Former Permanent Representative of Jamaica to the United Nations and Ambassador to the United States

S. D. Sylla
Administrateur,
Banque Internationale pour l'Afrique Occidentale

Adolph J. Warner
Former Vice-President and International Economist
Salomon Brothers (Investment Bank)

Maurice J. Williams
Executive Director
World Food Council

Joseph Yager
Senior Fellow
Foreign Policy Studies
The Brookings Institution

CONVENOR
Mostafa K. Tolba
Executive Director, UNEP

SECRETARIAT
Secretary: *Ruben P. Mendez*
Principal Officer, UNSO
Milena P. Roos
Research Economist, UNSO

UNEP CO-ORDINATORS
Yusuf J. Ahmad
Deputy Assistant Executive Director (Programme) UNEP

Noel Brown, Director, UNEP Liaison Office, New York

CONSULTANTS
H. E. Dregne, Director, International Center for Arid and Semi-Arid Land Studies, Lubbock, Texas, United States

M. Kassas
Professor of Plant Ecology, University of Cairo, Egypt

PART III

Feasibility Studies on and Detailed Modalities for Financing the Plan of Action to Combat Desertification (1981)

*Original text:
General Assembly Document
A/36/141 Annex
of 1 October 1981*

I INTRODUCTION

The present Part represents a third step towards the implementation of the 1977 Plan of Action to Combat Desertification by elaborating on the several previous studies on additional sources of financing contained in the previous parts of this book. The mandate to explore such additional sources originally grew out of a discussion at the United Nations Conference on Desertification (Nairobi, 1977). In its resolution 35/73 of 5 December 1980, the General Assembly, *inter alia*, requested the Secretary-General:

'To prepare, in consultation with the United Nations Environment Programme and with the assistance of similar groups of experts on the subjects concerned to be convened by the Executive Director of the Programme:

- '(i) Feasibility studies and concrete recommendations for the implementation of the additional means of financing deemed practicable by the Secretary-General, including those providing for a predictable flow of funds;
- '(ii) The detailed modalities of obtaining resources on a concessionary basis;
- (iii) A full feasibility study and working plan for the establishment of an independent operational financial corporation for the financing of desertification projects.'

In formulating these instructions, the General Assembly took account, *inter alia*, of the second study on financing the Plan of Action to Combat Desertification (Part II of this book).

That earlier study was not intended to carry beyond tentative evaluation of both feasibility and practicability of a number of specific finance mechanisms that had previously been put forward in the United Nations system. These evaluations found these two characteristics to differ greatly as between the various mechanisms. Based on this evaluation, the group of experts had then singled out certain of these mechanisms for more detailed treatment. One of these was a proposal for a special new institution to finance desertification and related projects. The group likewise considered in the study a variety of approaches to concessional financing, such as interest-rate subsidies and the possible blending of long-term funds raised in private capital markets with such subsidies or of guarantees. Such funds might then be channelled directly into certain desertification projects or, alternatively, go towards meeting the capital requirements of the institution referred to. Lastly, the study surveyed a list of specific sources of additional funds, drawing on existing pools of sovereign capital or tapping new types of revenues to provide a future automatic flow of funds. Two of these involve certain adaptations in the working of the international monetary system. Others are *de novo* sources. The proceeds from any of such additional sources could flow into projects directly or indirectly through the new institution or serve as backup for private financing.

The Executive Director of UNEP assigned the task of preparing the set of specific studies outlined in the instructions of the General Assembly to a core group of specialists selected from among those who had reviewed the previous study. These studies were once more reviewed by a group of high-level experts [whose names are listed in the annex. to this Part]. Their findings reflect the Assembly's judgement that all of the mechanisms described in the preceding paragraph were, in principle, feasible, but that a more concrete examination of the particular modalities of each was indicated. Accordingly, the studies review the economic feasibility of each mechanism, including its technical and legal aspects, as well as its administrative and practical aspects.

This study recognizes that, in practical terms, any additional funds raised for the objectives outlined in the Plan of Action are unlikely to remain confined to desertification control alone. Given the interrelationship between resources, environment, people and development,[1], this comment applies with equal force to all of the different mechanisms described below; desertification control forms just one aspects of economic development in general and of development in the related areas of energy, health, and agriculture in particular. The following considerations seem relevant in determining the proportion of additional resources that is reasonable to assign for anti-desertification purposes. Between 600–700 million people depend on desertification-prone areas for their livelihood; about 15 per cent of the world population [UNCOD (1977) figures, more recent estimates (UNEP, 1984) 850 million people, 20% of the world population]. The total crop-producing lands (rain-fed plus irrigated cultivated lands) prone to desertification are estimated at 200,180,000 hectares; about 13 per cent of the world croplands. The area of rangelands prone to desertification represents 25 per cent of the world rangelands and produces 10 per cent of the world livestock products. Accordingly, it appears that by various criteria, desertification control could reasonably qualify to receive between 10 per cent and 15 per cent of such additional resources as may become available to meet developmental, environmental and other financial needs of the international community.

II BACKGROUND

Desertification undermines the productivity of some 20 million hectares each year, a process that was dramatically highlighted by the disastrous

Sahel drought of 1968 to 1972 and the African famine of the 1980s. Desertification control thus forms an integral and increasingly indispensable part of the development process in the developing countries. The high cost and long gestation period of control programmes underscore the need for new sources and methods of financing. For the 20-year period to the year 2000, the 1980 study (Part II) put the total cost of the anti-desertification programmes in the developing countries at a medium range of $48 billion, with a low of $29.5 billion and a high of $79.6 billion, all in 1978 dollars (see Table 3, Part II, p. 68). This points to an average annual funding requirement of approximately $2.4 billion for that 20-year period. Against the average $2.4 billion annual expenditure implied by the medium total, the average financing needs was estimated at $1.8 billion a year net, after deducting $527 million of external assistance and $59 million of national expenditure concerned with desertification control that were provided, in the calendar year 1978, as follows:[2]

Member States of the Organization for Economic Co-operation and Development (OECD) plus international and regional financing agencies:	443 million
Bilateral and multilateral flows from the Organization of the Petroleum Exporting Countries (OPEC):	23 million
United Nations system other than World Bank and IFAD:	61 million
Total:	527 million
Add: National expenditures:	59 million
Grand Total:	$586 million

(a) The above totals are expressed in 1971 dollars and thus exclude the effect of price changes, as well as further environmental deterioration between that year and 1980.

(b) These estimates do not cover the cost of certain measures that may become necessary to implement the entire plan, the cost of which cannot be reliably estimated. Examples would be the costs associated with education and training, social change, and particular structural adjustments such as the stratification of livestock, i.e. by careful planning of the movements of livestock at various stages of their growth cycle to achieve the optimum distribution of herds throughout the different rainfall zones; intensified production and stockpiling of feeds and fodder; provision of alternative energy technologies; and the monitoring of desertification and related natural resource processes.

(c) The $1.8 billion net external assistance requirement is not split into local currency and foreign exchange components because neither local funding nor savings levels or revenue bases of the lowest-income countries affected would permit local currencies to be raised without external assistance. However, some part of such assistance could take the form of counterpart funds for food and other aid received *in rem* as outlined in the study contained in Part II of this book.

Subject to these qualifications, the funding requirement would rise from a starting level well below the average $1.8 billion, building up over time to well above that amount. This need not be seen as a negative factor, since the financing mechanisms evaluated in the following sections are unlikely to materialize, or to move forward toward implementation, simultaneously, or perhaps even in any predictable sequence. In part, this reflects the fact that these mechanisms vary with respect to the degree to which they meet the criteria of additionality and automaticity. Thus, these mechanisms may overlap internally in the sense that governmental agreement to any one of them may exclude consensus on some or all of the others. Similarly, an agreement on the part of governments to join in implementing some of the measures proposed here may come at the cost – expressed or implicit – of a slowdown in real terms, if not actual reductions, in other bilateral or other assistance programmes. Lastly, the automaticity, which represents an indispensable element of the type of long-term programming envisaged here, must likewise remain subject, even in a domestic context, to a measure of political and economic uncertainty over the entire 20-year planning span. Yet it is imperative that a start toward this objective should not again be delayed until new unforeseen contingencies once again overtake the initiatives needed to begin funding the Plan of Action.

Nevertheless, none of these comments is inconsistent with the General Assembly's judgement that all of the mechanisms discussed in this report are, or could be rendered, feasible. To be sure, they remain subject – albeit to varying degrees – to the more detailed examination of their technical practicability, including the legal and administrative aspects as elaborated in the individual feasibility studies that follow. Taken together, these findings actually highlight the critical point that this four-year sequence of studies has now reached, and the implications which all of them carry for meeting the goals of the Third United Nations Development Decade. Just because the nature of desertification projects and the returns associated with them render their financing difficult, such a final breakthrough toward originating, and success in implementing, new mechanisms would constitute a new signpost for determining the development patterns of the Third Development Decade.

The present study designated six proposals outlined in the earlier study contained in Part II as falling within the limits of appropriate degrees of feasibility and thus proper subjects for the following detailed studies. While all meet the criteria of additionality and – albeit to varying degrees – of predictability as well, they do differ in certain important respects. These differences must be kept in mind while evaluating the concrete recommendations for their implementation.

In considering the implementation of the additional means of financing, it is equally important to bear in mind that certain broad perceptions have emerged. In the first place, the international community appears to be agreed on the importance of a number of plans of action to meet critical concerns, one of the foremost among these being the PACD. The second perception relates to the fact that the countries and regions most seriously affected by desertification are precisely those that are least able to cope with the magnitude of the problem. Not only are additional resources required but those resources must clearly be available on a concessionary basis. Thirdly, the environmental concerns, such as the spread of desertification, are not only grave but extremely urgent. Finally, for a number of reasons, such as the need for a measure of automaticity and predictability in the flow of resources and the fact that desertification is not and cannot be confined to national boundaries, international measures are to be preferred to bilateral approaches.

III   FEASIBILITY STUDIES AND CONCRETE RECOMMENDATIONS FOR THE IMPLEMENTATION OF THE ADDITIONAL MEANS OF FINANCING

A   Generalized trade taxes

The general concept of funding economic development through taxes on international trade in goods and services has formed part of the literature on the subject both within and outside and United Nations system for some time past. The first formal request for a study of the practical use of such taxes by any United Nations body originated with the PACD.[3] In the following year, a study published by the Brookings Institution[4] examined the subject in some depth, together with other additional measures for financing, some of which are addressed elsewhere in the present study. In 1979, the Brandt Commission Report included taxes on trade among the measures it recommended as useful for enlarging the flow of official development finance.[5] In turn, they were listed, together with other variant of international taxation, in last year's comprehensive inventory of all additional means of financing proposed in the United Nations System. The trade tax concept was also pursued in UNCTAD and discussed at the Trade and Development Board's Nineth Session (first part) held at Geneva in July 1980.[6]

Most of these studies, *inter alia*, try to evaluate the relative merits of two alternative approaches to such taxation: either a gross sales tax levied world-wide on the broadest possible mix of goods and services – the latter including tourism and transportation, along with financial services (such as investment income and revenues from banking and insurance), all with the least possible amount of exemptions; or, alternatively, a more narrowly-based tax, levied on specific categories of goods, and often structured so as to take into account its relative impact on the contributing countries' income and wealth. The following discussion covers both types of levy and examines the feasibility of applying them both to goods and to services. On the other hand, the discussion limits itself to taxes intended primarily to raise revenue, rather than those designed mainly for their penalty effect – such as taxes on polluters of the environment, or taxes on shipments of military goods – or for their disincentive value, even though such effects are, to varying degrees, inherent in either type of tax.

In evaluating the economic and financial significance of any one or more of these various forms of taxation, the different modalities described below have been examined from the viewpoint of their revenue potential, ease of administration, fairness, and economic effects. In turn, these criteria have served as the basis of recommendations. As in other sections of this study, and given the complexities of collection and compilation, not all of the underlying statistical data are available in up-to-date form, nor are they always as comprehensive and reliable as might be desired. They are, nevertheless, considered to be fully adequate within the context of the mandate to make detailed, concrete, and action-oriented recommendations.

As brought out in Chapter I of Part II of this book, discussion need not be confined to the taxation of merchandise trade alone, it may be extended to taxation of invisibles: levies on international investment, air travel, and freight transport, all items that had meanwhile also been suggested by the Brandt Commission Report. In addition, the earlier Brookings Institution study had more systematically examined the merits and possible modalities of taxing invisibles, in which category it also included tourism. However, it was considered that, because of administrative and other factual reasons, an international trade tax on merchandise trade should be considered (in the first instance), notwithstanding the possibility of considering the taxation of invisibles as the system matures.

Keeping within the terms of the mandate of the study, which calls for a feasibility study of international taxation of all trade flows, levies on international investment (presumably interest and dividends) or on surpluses in foreign trade have not been considered, as these do not fall within that definition. The feasibility of taxing service transactions, both in absolute terms and relative to levies against merchandise trade, however, is being addressed below.

It is recognized that, by some measures, the value of service transactions may have grown faster than that of merchandise trade – particularly if the latter's value is taken net of trade in energy resources. Even so, estimated in the roughest of terms, service transactions probably amount to no more than one-fifth of the current $2,000 billion global value of merchandise trade. But some of these services, such as banking and insurance charges, are largely ancillary costs of merchandise trade and thus likely to be passed on to the final purchaser in one form or another. Thus, any taxable revenues from insurance and freight would be included if taxes on merchandise trade were imposed at the c.i.f. price (cost, insurance and freight) instead of the price free-on-board (f.o.b.). This would permit a lower level of administrative expense than that involved in a separate collection process; most proposals for the taxation of merchandise trade therefore use c.i.f. values.

Transportation and communications across national borders – not infrequently operated by public bodies – are other categories of services that are inherently difficult to capture for taxation except at considerable administrative costs. This would be less true for international air travel, which is generally associated with either business or tourist travel and thus often considered as falling in the luxury category, as distinguished from such 'essential' services as insurance and freight. However, fairness would call for a good deal of differentiation – for instance, between tourism expenditures in wealthy countries and those in low-income developing countries aspiring to build up their income from that source. To avoid such taxation becoming counter-productive would require further modification, with all the attendant administrative complexities, by making special allowances for 'essential' components like landing fees and airport or marine tender services, which could be considered as luxuries only from the viewpoint of their end use. As a final administrative barrier, many types of services are not generally subject to import duties or tariffs, so that the regular customs agencies could not handle the collection of taxes. A similar argument applies, *a fortiori*, to service payments for commissions, royalties, and management or copyright fees where the imposition of a tax would most likely cause widespread circumvention and evasion. In view of these considerations, it is considered that the proposed trade tax should remain limited to merchandise for the purposes under discussion here. This conclusion seems to have been shared by the UNCTAD group, which likewise confined its recommendations to the taxation of merchandise transactions.

It is tempting to view international taxation as an extension of domestic taxation and to treat it as a simple surcharge – or so-called 'shadow tax' – levied along with a country's other taxes, whether direct or indirect, on behalf of an international organization. Such taxation would, of course, constitute an entirely voluntary scheme, with no assurance of permanence, even if it were possible to obtain a sufficient measure of agreement. To cement such consensus, once established, into a more binding form of agreement would require an international treaty for a trade tax administered by a special body. While either approach presupposes an act of will on the part of governments, the latter procedure promises somewhat higher predictability, once it has been ratified by a sufficient majority of countries and approved by their national legislatures. Moreover, legislatures will usually find it easier to accept a new tax levied for specific purposes rather than to let the international community pre-empt their traditional sources of domestic revenue. This consideration would seem to rule out any kind of surcharge on existing taxes, whether related to income (direct) or to expenditures (indirect). World-wide, more than twice as much revenue is raised through indirect taxation as through direct taxation; and that proportion is on the rise as more governments are adopting simple sales or value-added taxes, while reducing income taxes.

With world trade currently running at about $2,000 billion a year and likely to reach $2,500 billion by 1984, the principal advantage of such a tax is its very large base. [Recent estimates by UNCTAD show the figures for world trade in 1986 could well be in the region of $2,159 million]. Provided that the tax is levied on essentially all commercial trade and on all trading nations, even as low a rate as 0.1 per cent would yield $2.0 billion a year currently and perhaps $3.0 billion by 1987. Assuming that up to 10 per cent of this revenue might become available to combat desertification, we estimate that the tax could yield an average of over $250 million a year through the 1980s. It would thus contribute much more massive amounts than even the maxima likely to be obtained from either IMF Trust Fund reflows or SDR allocations, and would presumably continue in effect, i.e. provide greater predictability. At the same time, the tax rate of 0.1 per cent is too small to be considered as threatening to reduce the volume of trade through import substitution, or exert any sufficiently visible pressure to raise consumers' inflationary expectations to any meaningful extent. The initial discussions of the requisite legislative measures should overcome such objections by specific reference to the purposes for which the tax is earmarked (such as desertification control or other development programmes). An international agreement directing the body administering the tax to exercise these functions would, *inter alia*, specify the currencies which would be acceptable from national authorities. Presumably, these would consist of a number of freely-convertible currencies, such as the five currencies constituting the SDR, together with a smaller proportion of local currencies to be used for expenditure in the countries concerned. Similar arrangements would govern trade denominated in bilateral clearing currencies.

Lastly, the availability of a virtually world-wide network of national customs authorities with common classification standards facilitates the collection of what is, in essence, a surtax on both dutiable and non-dutiable trade. Since customs authorities generally value even non-dutiable goods for statistical purposes, their capture for purposes of the levy would generally not represent a problem. While the tax would normally be collected from imports (c.i.f.) at ports of entry, there may be some advantage in alternative collection from exports at ports of shipment in order to provide greater assurance of collection in cases where certain importing countries may balk at meeting their agreed obligations.

Notwithstanding this powerful array of advantages, the parallel with national sales taxes also suggests a number of negative elements that necessarily characterize any international trade tax. All broad-gauged consumption taxes are unavoidably regressive in that they weight most heavily on low-income consumers, whose purchases of necessities claim higher proportions of their income than do those of the more affluent. In the same way, it is often the lower-income (especially smaller) country that shows the greater dependence on imports, while larger and more populous countries tend to be less import-dependent, regardless of their income levels, simply because their greater size and resources provide them with a greater measure of self-sufficiency. In actual fact, differences in population and size, more than in income and wealth, explain most of the discriminatory effect of an import surcharge. As Table 1 demonstrates, imports of the larger countries tend to absorb lower percentage shares of their total income than do imports of smaller, less populous countries, regardless of the absolute level of their income per capita. The table compares the relevant percentage shares of five of the world's most populous countries with those of ten of the less populous ones.

These data clearly show the discriminatory effect of taxing various countries' imports at a uniform rate. Taking that rate as 0.1 per cent of the c.i.f. import value, the United States would contribute 0.01 per cent of its GNP and the Soviet Union 0.00 per cent; these low ratios compare with a range of 0.03 per cent for Trinidad and Tobago to 0.07 per cent for Liberia. Thus the latter country would be liable for seven times the GNP

Table 1

Country	Population (millions)	Imports (c.i.f.) GNP
	(1979 data, except as noted)	
China (1977)	950	2.0 per cent
India (1977)	626	7.1 per cent
Union of Soviet Socialist Republics	265	5.7 per cent
United States of America	220	10.4 per cent
Brazil	119	9.6 per cent
Sudan	17	16.0 per cent
Kenya	15	32.6 per cent
Sri Lanka	15	45.7 per cent
Netherlands	14	52.3 per cent
Sweden	8	32.8 per cent
Norway	4	42.8 per cent
Liberia	1.7	73.2 per cent
Mauritania	1.5	61.8 per cent
Trinidad and Tobago	1.0	31.8 per cent
Gabon	0.5	50.8 per cent

percentage contributed by the United States, and Mauritania for ten times the Soviet Union's. Because this would not be found acceptable, adjustments would be required. The tax levy on merchandise could be corrected by a variable representing the degree of dependency of each country on international trade so as to achieve greater equity between the contribution require from countries heavily dependent thereon and from those with a greater degree of self-sufficiency. Such adjustment may, for instance, introduce one or more ceilings which limit the rate of tax to a maximum percentage of GNP for smaller developing countries with low per capital incomes. Such adjustment would not provide a significant degree of relief, however, in an environment where these low-income developing countries' imports have been declining from 2.1 per cent of world-wide imports in 1970 to only 1.5 per cent in 1977 – a larger decline than that registered by all developing countries relative to the value of global imports. Instead, population data must be brought into the equation, although, as indicated above, not necessarily in the form of per capita GNP.

One possible formula for effecting this adjustment was suggested by the 1978 study by the Brookings Institution,[7] which had the small, more trade-dependent developing countries retain a part of their tax collections, the rate of retention varying inversely to each country's population. That study found that imports tended to average 21 per cent of GNP for medium-sized countries (30 million population), 10 per cent for large ones (200 million population), but 43 per cent for small countries (1 million). Overall, the effect of applying the Brookings formula was to reduce by one half this group of countries' tax burden, expressed as a percentage of social product. By way of example, where a 1 per cent tax on all imports would have taken 0.16 per cent of all countries' combined GNP, adjusting the tax to 2 per cent for the non-retainers and applying the retention formula to the smaller country group would reduce the latter's global burden to 0.075 per cent of its combined GNP, without significantly changing the total tax collected.

Even this adjustment would still leave substantial inequities in the distribution of the burden between the retaining and the 'normal', i.e. non-retaining countries as a group, as well as between countries individually. The formula does not extend to countries with the largest populations where import/GNP ratios of 7.5 per cent or less prevail; the breakpoint occurs at a population level of 150 million. Thus, applying the alternative formula, with the tax rate doubled to 2 per cent, would also double the contributions of the United States (from 12.9 per cent to 25.5 per cent of total tax collections), India (from 0.6 per cent to 1.2 per cent), China (from 0.6 per cent to 1.2 per cent), and the Soviet Union (from 3.3 per cent to 6.5 per cent). The benefits would be distributed even more randomly as between higher- and lower-income countries. Exclusive of the doubling of the United States' contribution, the formula would actually reduce that of all other countries with income per capita of $2,000 and over by more than 14 per cent and would cut that of the oil-exporting countries by almost one-third. As a result, the oil exporters' contribution, which had equalled 1/16th that of the entire $2,000 and over group, would have shrunk to less than 1/25th – clearly not a pattern of burden-sharing likely to find ready acceptance then or now.

One main reason for these inequities lies in the choice of imports as a basis for the trade tax, without reference to countries' trade balances – plainly a more useful indication of their ability to contribute. Obviously, it would be logical to expect deficit countries to be more seriously affected by an import surcharge because import elasticities are lower than those for exports, especially in low-income countries with a large proportion of essentials in their import mix.[8] The formula does not make this distinction, even though that proportion has risen faster for the lower-income countries than it has for the others. Between 1960 and 1977, the proportion of total imports absorbed by food and fuel rose from 33 per cent to 35 per cent for the industrial countries, but from 29 per cent to the same 35 per cent ratio for the lower-income ones. Given the further increases in the costs of energy imports since 1977, the trend has accelerated to the detriment of the latter group of countries.

It may well be that these and similar inequities are inherent in any rate differential or rebate scheme, whether the amount of rebate is made more directly income-sensitive than in the Brookings version or not. Following that scheme on the basis of available data for 1979, but applying a uniform 0.15 per cent rate of tax without rebating and a 0.3 per cent rate, rebated according to the Brookings formula, would not support a materially different conclusion. Nor, as explained earlier, would these inequities be significantly reduced by apportioning the total tax burden in direct proportion to per capita incomes, except for reducing modestly the percentage gap between the industrial countries' contribution and that of the oil-exporting countries. On balance, simply exempting the lowest-income countries, regardless of their population size, would be no more inequitable and would simplify administration. This approach has the further advantage of not requiring frequent adjustments, as countries' circumstances change, even though it would tend to violate the principle of universality. Moreover, it would be considerably less cumbersome than to apply lower tax rates to those countries' essential imports, like food, medicines, and energy resources. The *prima facie* appearance of feasibility, as referred to in Part II of this book, thus comes down to the recognition that the decisions required for their implementation involve judgements that remain controversial or political, rather than technical considerations alone. The resolution of these issues involves the acceptance of some of the inequities referred to, mainly by those countries best able to agree to do so, as an expression of their political will to marshall truly additional real resources for the most urgent demands of long-term development, without reducing other contributions to ODA. Once such a consensus is established, the legal and technical aspects of their implementation will *ipso facto* become more easily manageable.

The legal foundation for any international undertaking – whether revenue-raising or not – of the size and complexity involved in taxing world trade would be a basic treaty or convention to be negotiated and ratified by an agreed minimum number of countries in order to permit it to enter into force. The treaty would operate through a permanent international treasury created to carry out its provisions, including, *inter alia*:

(a) organizing the working relationships with sovereign governments, and administering tax collections, by means of rules and regulations;
(b) supervising the operations of a trust fund that would receive and disburse tax collections and possibly borrow against expected future receipts;
(c) settling disputes by arbitration or otherwise, and using sanctions where required to enforce collections.

The composition of that body's management and policy-making apparatus should, in principle, be structured so as to reflect the relative importance of each signatory power's expected revenue contribution. In practice, countries may insist on this as a pre-condition for their ratification of the treaty. At the operating level measures should be taken to ensure the democratic participation of all Member States in the decision-making process. The adjudication of disputes and imposition of sanctions may call for the creation of a separate panel or court.

Acting, in effect, as agents of this newly-created body, national customs agencies will collect the trade tax directly from importing firms and individuals, based on c.i.f. values (thus effectively taxing insurance and freight along with the value of the merchandise). Taxes on dutiable goods would thus be collected as a surcharge on existing Standard International Trade Classification (SITC) categories and tariff schedules. The tax would apply to all non-dutiable goods, but would exclude commercial samples; special consideration should be given to transit trade, imports into free ports and free trade zones, as well as otherwise exempted trans-border processing transactions. In countries that refund local value-added taxes to exporters, such taxes would be excluded from invoice values in computing the amount of the trade tax, as would any corresponding import-equalization tax imposed by importing countries.

Special discretion will be required in areas where such apparently technical decisions actually touch upon issues involving trade policy. Among these, any double taxation of re-exports is of particular concern to smaller developing countries and territories like Singapore, Hong Kong and Trinidad and Tobago, whose high import/GNP ratios would otherwise subject them to unreasonably high tax burdens even under the Brookings tax retention scheme. Seen more broadly, the growth of world trade over recent decades owes much to liberalization measures such as the expansion of customs unions, the extension of most-favoured-nations (MFN) treatment, and similar arrangements. Without in any way diminishing the benefits of such arrangements, these must remain neutral for purposes of the trade tax, on the principle that universality requires the taxation of all trade that clears customs borders, including, for instance, those between the members of the European Economic Community and the European Free Trade Association, and between the Federal Republic of Germany and the German Democratic Republic.[9] In the Eastern trading area, as well as in certain other jurisdictions where state trading prevails, equally tax-neutral provisions must be made to apply in the valuation of goods, including those invoiced in clearing currencies, or exchanged in barter transactions which possibly involve non-market pricing. In large part, such technicalities will likewise fall within the purview of the newly-created agency, but provisions designed to address some of the problems mentioned above may well have to be incorporated in the treaty itself.

In considering the economic impact of any general trade tax, its advantages – meaning its broad base and the low tax rates at which it still yields large revenues – carry with them certain disadvantages. Thus, the question of whether and how much of the tax would ultimately be borne by the importer/buyer, or how much of it that importer can shift back to the exporter/seller, cannot be answered validly for such a large variety of goods without detailed demand and supply data that would, together with historic prices, permit the establishment of demand/supply and price elasticities. And even with such data in hand, attempts to express these elasticities numerically would not provide any meaningful conclusions at rates of tax running as low as 0.1 per cent to 0.4 per cent. The impact on any individual (buying) country depends similarly on the elasticities attaching to its particular mix of imports. Countries with a high proportion of low-elasticity essentials, like food and fuels, in its total imports will, by definition, have to absorb larger parts – if not all – of the tax, at least in the short run, because of their dependence on imports. Where possible, import substitution through local production might play a role in the longer run, but seems extremely unlikely to be triggered by tax levels of the order of magnitude proposed here. The same *de minimis* judgement applies to the question of the impact of the tax on cost and price levels, i.e. its inflationary impact, particularly in countries unable to shift back the tax burden. Even after allowing for transmission of this cost in subsequent processing stages, for raw materials, and commercial (resale) levels, including the attendant profits, it is difficult to see a coefficient as high as 1.5 applying to the basic import tax, i.e. an effective impact of 0.15 per cent for a 0.1 per cent tax rate, after including all the secondary price effects resultant. This would mean a one-time inflationary impact of 0.075 per cent in a country with an import/GNP ratio as high as 50 per cent but would equal only 0.0125 per cent worldwide. Given the commanding need for additional sources of development funds with a high rate of predictability, this cost cannot be considered as excessive.

The revenue-raising potential of an international trade tax is thus high. Moreover, once adopted, the tax would afford a considerable degree of automaticity, in that a country would have to withdraw from the treaty for its contribution to cease. Based on the assumption that no more than 10 per cent of the additional revenues created by the tax would be devoted to desertification control measures, tabulation shows a flow of funds that would produce $200 million (at 0.1 per cent) on last year's trade, rising by 50 per cent to $300 million by 1987 (see Table 2). This reflects an average annual gain of only 6 per cent in nominal terms – an estimate tending toward the lower end of current expectations.

Table 2

	(in $US billions)								
	1982	1983	1984	1985	1986	1987	1988	1989	1990
Trade tax at 0.1 per cent	2.25	2.38	2.52	2.67	2.84	3.00	3.18	3.37	3.57
Trust Fund from IMF gold sales	—	—	—	—	0.29	0.29	0.29	0.29	—
SDR-link	—	0.92	0.92	0.92	0.92	0.92	0.92^a	0.92^a	0.92^a
Satellite parking fees	—	—	—	0.03^a	0.04^a	0.05^a	0.07^a	0.08^a	0.10^a

[a] Indicates figures that are more speculative than others.

These sums could be channelled directly into anti-desertification projects. Alternatively, some modest fraction could be used to leverage the borrowing capacity of the institution organized to finance such projects, as outlined in section IV.

B IMF gold sales and Trust Fund reflows

It is unnecessary for present purposes to recapitulate the circumstances under which the IMF came to divest itself of 50 million ounces of gold or one-third of its original total gold stock of 150 million ounces, except to note that one-half of this fraction, namely 25 million ounces, was auctioned publicly over a four-year period beginning in 1976; the other half was restituted to the original contributors. These gold auctions produced total profits in excess of book value of $4.64 billion, of which $1.29 billion was distributed directly to the 104 developing country members of IMF with nearly 28 per cent of total quotas. The remaining gold sale profits of $3.35 billion, corresponding with investment income, etc., to SDR 2.9 billion,[10] was transferred to a Trust Fund established in May 1976 for the benefit of certain developing countries (see following paragraph) and which began operations in 1978.

Trust Fund loans have a maturity of 10 years and a grace period of five years so that, while repayments can commence on some loans in 1983, the first repayments on the last instalment of Trust Fund loans fall due only in 1986. It is necessary, therefore, to examine what part of the repayment of Trust Fund loans could be made available for anti-desertification purposes in the light of two decisions made by the Fund's Executive Board pre-empting the use of these reflows for certain other purposes.[11] In the first place, SDR 750 million of Trust Fund reflows have been earmarked for a Subsidy Account intended to reduce the cost of using the IMF's Supplementary Financing Facility (the Witteveen facility) for its low-income members. In the second place, another SDR 1.5 billion has been set aside to provide balance-of-payments assistance on concessional terms on a uniform basis to low-income developing members in need of such assistance under arrangements broadly similar to those of the Trust Fund. The Trust Fund was originally established to provide additional balance-of-payments assistance on concessional terms to eligible developing countries that qualified for assistance by carrying out programmes of balance-of-payments adjustment. Unless, therefore, countries with desertification problems have gone through the same rigorous discipline necessary to benefit from the Supplementary Financing Facility, or have accepted the considerably milder degree of conditionality required to benefit from Trust Fund loans, they cannot expect to benefit from these two sources of Trust Fund reflows totalling SDR 2.25 billion. In any event, no part of this amount can be available in terms of existing Fund Board decisions for interest subsidy purposes against which a multiple amount of Third Window-type loans can be raised under guarantee.

However, a residual amount due under Trust Fund repayment procedures does remain as yet unpre-empted by any prior Fund Board decision. The rough SDR equivalent of the original $3.35 billion transferred to the Trust Fund – about SDR 2.9 billion – had been lent out when the Trust Fund was wound up, following the disbursement of two final loans on 30 April 1981. This still leaves a difference, as yet unallocated by IMF, between the SDR 2.9 billion lent under the Trust Fund and the SDR 2.25 billion of reflows already earmarked, as described by the Fund Board, for essentially IMF purposes. This amounts to SDR 650 million in all. However, commencing no later than 1986, when payments begin on the last instalment of Trust Fund loans, this amount with additional interest income accruing to it over the period 1986–89 could amount in round figures to SDR 1 billion, which will be available to provide assistance to low-income developing members in accordance with the second section of subsection 12.f (ii) of the Fund's Articles of Agreement under a decision of the Fund to be taken not later than 30 June 1988. This subsection of the IMF Articles specifies that 'balance-of-payments assistance may be made available on special terms to developing members in difficult circumstances and for this purpose the Fund shall take into account the level of per capita income'. The sub-section also specifies that action under it should be 'consistent with the purposes of the Fund'. These purposes are, of course, defined in article 1 of the Fund and, in the light of evolving thinking concerning the nature of the balance-of-payments adjustment process in developing countries, there seems every reason to suppose that the SDR 1 billion not so far earmarked by the Fund Board can be made available, *inter alia*, for anti-desertification purposes. Given the time pattern of repayment of Trust Fund loans, SDR 1 billion can be expected to accrue in four annual instalments of SDR 250 million each during each of the years 1986 to 1989.

As will be argued in the next section C, the implementation of a link between SDR and development finance, could, on conservative assumptions about the scale of annual SDR allocation and the proportion of SDR that developed countries would agree to forego, be expected to yield over the fourth basic period 1982–86 some SDR 800 million. On the assumption that 10 per cent of this amount can be earmarked for anti-desertification purposes, this yields annual interest subsidy amounts of SDR 80 million. On the assumption that SDR allocations will continue at least on the same scale during the subsequent basic period covering the years 1987–91, a total of SDR 1.05 billion would become available for overall development purposes annually from 1986, when Trust Fund reflows are also added in. On the previous assumption that no more than 10 per cent of this amount could reasonably be earmarked for anti-desertification purposes for each of the years 1986 to 1989, SDR 105 million is made available, as compared with SDR 80 million annually in 1982–1985, which can be used for purposes of subsidizing concessionary loans to be raised under guarantee (see Table 3).

Action within the provision of article V, section 12, of the Fund's Articles of Agreement requires a special 85 per cent majority of the Fund Board. What makes it probable that the relevant majorities can be mustered stems from evolving trends in both the policies of IMF and the World Bank. While IMF has been steadily extending the maturities of its lending either through an extension of the maturity of the Extended Fund Facility to 10 years or through the device of successive stand-by agreements broadly agreed to in advance, the World Bank has recently

Table 3

	(in million SDRs)							
	1982	1983	1984	1985	1986	1987	1988	1989
SDR-link proceeds	800	800	800	800	800	800	800	800
Trust Fund reflows	—	—	—	—	250	250	250	250
Total resources available through IMF mechanisms	800	800	800	800	1,050	1,050	1,050	1,050
Amount available for anti-desertification purposes	80	80	80	80	105	105	105	105

introduced its structural adjustment facility also for extending programme loans to countries. Both sorts of evolution reflect the recognition that the conventional three- to five-year time-period previously thought relevant for balance-of-payments adjustment, and the concentration, in the Bank's case, on project as compared to programme loans, have the effect not of promoting but of frustrating the adjustment process and of imposing avoidable social and economic costs on the borrowing country. Central to both sorts of evolution is the recognition that, in the case of a developing country, exchange rate charges *per se*, in the absence of more purposeful steps of a planning character involving both import substitution and export promotion activities, both of which take time for the necessary capacities to be installed, are not in themselves likely to suffice. These considerations apply *a fortiori* to the low-income country saddled not only with the task of development but with the task of combating desertification as well.

Increasing recognition of the necessity of bridging the gap between adjustment and development financing is now in process.[12] It is this shift of opinion about all which makes it reasonable to suppose that IMF can allocate some portion of its resources to an anti-desertification institution whose purposes relate much more closely to longer term development. There would appear to be nothing in the purposes listed in article 1 of the IMF Articles of Agreement that militates against such course of action as indeed against the establishment of a link between SDR and development finance.

Future sales programme

The group reiterated the necessity to promote the common objective of making the SDR the preferred international reserve medium by phasing out gold and national reserve currencies and the potential it may hold providing assistance to developing countries. Likewise, it felt that an effort should be made to adjust the distribution of international liquidity in favour of developing countries in the light of the massive potential transfer to the major gold-holding countries of some 400 billion dollars as a result of the effective revaluation of their reserve asset gold holdings.

Towards this end, four-fifths of the remaining 103 million ounces of gold held by IMF would be auctioned or otherwise disposed of over a 10- to 15-year period, and that portion of the profits not directly distributed to developing countries earmarked for new international assistance programmes for developing countries, except as noted in the following paragraph.

Out of these proceeds, one-tenth could be earmarked for financing desertification control programmes. Subject to the requisite approval by a qualified voting majority, this could probably be done within the present Articles of Agreement of the IMF. A total amount of the order of $3 billion would thus be released over this 10- to 15-year period out of which $2.2 billion would represent the transfer of the other members of the international community to the anti-desertification programmes in the developing countries. Developing countries in a financial position to do so would be expected to announce that they would not avail themselves of credit facilities or funds made available through such transfers.

Alternatively, and if necessary, subject to appropriate amendment to the Articles of Agreement of the IMF, the above-mentioned portion of the profits from these new IMF gold sales could be kept by the IMF in a segregated trust fund and invested so as to permit its use as collateral for guaranteed loans which the proposed corporation would raise on the international financial markets, whose terms and conditions would then be subsidized either from the other sources envisaged in this report or from the income from the trust fund.

To assure the maximum returns from any future sales programme, such sales should again be spread over time in line with the market's absorptive capacity. Sales would have to be predicated on projections of such capacity at given price levels: it will be recalled that the amount of monthly Fund sales had to be reduced from 525,000 ounces to 470,000 ounces, following the United States series of auction sales.

The wide fluctuations of the gold price would suggest that future sales strategies should focus on methods for capitalizing on such price volatility by: (a) pre-determining price levels that would trigger new auction sales; and (b) authorizing alternative means of disposing of gold, including sales on the open market and perhaps the writing of call options whenever market conditions would make such alternatives more attractive. (The latter represents virtually the only way of extracting a return from the Fund's gold stocks regardless of whether it actually disposes of any part of such stocks.)

Yet another alternative would be to follow the Brandt Commission recommendation to use the IMF gold as collateral for borrowing. This would not require that any part of the IMF gold be sold, so that it can remain within the IMF to serve this purpose. Still, the use of IMF gold as collateral has the complication of requiring possible time-consuming amendments to IMF articles whereas, as mentioned, the gold sales procedure can probably be accommodated within the present articles, thus requiring a qualified majority.

C Link between SDR and development finance

As is well known, SDR are currently allocated in proportion to the quotas of Fund members so that, roughly speaking, 26 per cent of any such allocation now accrues to the non-oil developing countries, 10 per cent to oil-exporting developing countries and the balance to the developed countries. The proposal to link SDR with development finance has a long-standing history, going back to the Stamp Plan of the 1950s, under which all SDR-type deliberately-created international liquidity – there being no SDR as such as that stage – would accrue to developing countries. Since then, the current of established opinion has veered in favour of having (as emphasized in 1969 by the Pearson Commission) 'the developed countries relinquish a part of their quotas of the new reserve medium (SDR) in favour of the less developed countries', on the grounds that 'there are strong reasons of simplicity and equity' for so doing 'when the scale of the issue of SDR has been decided on appropriate grounds',[13] namely those relating to world liquidity and not the needs of development finance. The Pearson Commission proposal related to a form of link where SDR were channelled to the IDA and re-lent to developing countries. Since that time, however, various alternative mechanisms for implementing the link were developed, especially during the discussions of the Group of 20.

These may be illustrated as comprising the following schemes:

(a) The allocation of SDR directly to development financing institutions (DFI);
(b) The allocation of SDR directly to developing countries entailing a larger share in SDR allocations than their share in fund quotas;
(c) An increase in the share of developing countries in Fund Quotas and hence in SDR allocations;
(d) An 'indirect link' under which SDR allocations would remain proportionate to Fund Quotas, but would be accomplished by an agreement among donor countries to transfer a predetermined portion of their SDR allocations or the equivalent in currencies, to DFI;
(e) A link between grants to DFI or developing countries directly and the establishment of a 'Reserve Substitution Facility' (a precursor of the more recent substitution account proposal).

Since that time more recent discussion within the IMF has brought to surface two new possibilities, labelled F and G, which are briefly described as follows:

Scheme F: A link to Fund-supported programmes having the purpose of reinforcing the incentive for adjustment provided by the availability of conditional resources through the Fund. One variant of this scheme would provide additional resources in the form of SDR to developing countries that would qualify for IMF stand-buys under the normal conditionality rules.

Scheme G: An interest subsidy link under which SDR would be distributed for the purposes of subsidizing payments of charges to the Fund by developing countries with outstanding purchases from the Fund. Under this scheme, SDR could play an identical role to Trust Fund reflows in subsidizing interest-rates payable by the borrower. With SDR interest-rates now at market levels, an illustrative calculation made by the Fund suggests that as much as 1 billion SDR would be required to reduce interest payments to low income countries by 3 percentage points under the Supplementary Financing Facility. As explained below, such an SDR allocation for interest subsidy purposes, if transferred as an indirect link by the original recipient together with ancillary arrangements for shifting charges on to the facility, makes the transfer tantamount to a line of interest-bearing credit with which to meet interest cost. This cost could be avoided if again, as explained below, the amounts take the form of counterpart national currencies rather than SDRs as such.

In the course of 1980, the question of the link has been revived in Fund Board discussion both as part of the IMF response to the programme of Action of the Group of 14 tabled at the 1979 annual meeting in Belgrade and as part of the consideration of the Brandt Commission Report, which came out in favour of the link. The discussion within the IMF has also been influenced by changing economic circumstances which perhaps give more grounds for hope concerning the establishment of a link now than at any previous time in the history of the discussion of this question. The principal argument favouring the establishment of a link now derives from the expected persistence of developing country deficits as a counterpart of current account surpluses elsewhere in the system, the difficulties attending the smooth recycling of oil-producing countries' surpluses by private banks at the present time, and the difficulty of looking to increased official development assistance out of government budgets. If, therefore, extra-budgetary sources of finance can be found to meet developing country deficits at the present time, this can serve to offset the recessionary influence deriving from sluggish aid flows and any weakening of the recycling processes by private banks. Given the constraints affecting other available sources of finance, the importance of an SDR link becomes obvious in this context.

Although the SDR has now lost its previous element of concessionality by the decision to charge interest at market rates, an allocation still has the effect of extending a perpetual line of credit at market rates to developing countries which are net users of SDR, but will by the same token confer market rates of return on countries who increase their net holdings of SDR. This results from the present Articles of Agreement of the IMF. A paticipant receiving an allocation of SDR assumes, *inter alia*, an obligation to pay charges on that allocation equivalent today to market rates of interest. As long as it holds the SDR allocated to it, it will also receive interest on these holdings. Since the rates at which charges and interest accrue are equal, the allocation of SDR *per se* imposes no net cost upon participants. It is only when a participant reduces its holdings of SDR below its net cumulative allocation for whatever reason, whether by transferring them to a development institution or by settling its own payments deficits, that it incurs a net interest charge; it loses the right to receive interest on the SDR it uses while retaining the financial obligation to pay the charges attached to their allocation. It is for this reason that, as described above, the SDR allocation is tantamount to the opening of a line of credit which does not involve a net interest cost until the credit is actually used. Conversely, as mentioned, a participant adding to its net cumulative allocated by 'accepting' SDR obtains a net return that is now equivalent to a market rate of interest.

The effect of a link, whichever form is adopted, is to increase the perpetual line of credit being extended to developing countries who benefit, given any particular global total of SDR allocations, as compared with the situation where SDR are allocated in the normal way in proportion to Fund quotas. This increase represents the SDR amounts foregone by the developed countries in implementing the link. Such an enhanced line of credit will of course facilitate the access of developing countries to normal capital markets and therefore enhance their capacity to finance balance-of-payments deficits. At the same time, the market rates of interest attaching to SDR will permit confidence to be maintained in the asset so that the developed countries which will be foregoing some part of their normal SDR allocation as a result of the link will continue to have an inducement to earn and 'accept' SDR.

The principal question that arises in the current context concerns which mode of implementing the link is most suitable from the standpoint of providing an interest subsidy element in an anti-desertification programme. The principal consideration concerns the form of link which can be implemented most speedily without an amendment of IMF articles. This consideration points unambiguously in favour of the type (d) link described earlier, the so-called 'indirect link'.

A second consideration relates to whether such an indirect link is to be implemented by transferring SDR allocations as such to an anti-desertification institution or the equivalent in currencies. As already explained, the effect of an 'indirect link' is that countries foregoing SDR to an anti-desertification institution will continue to bear the financial obligation to pay charges attaching to the original SDR allocation. This obligation can, it is true, be passed on to the anti-desertification institution by agreement, but this can only have the effect of diminishing the resources available to such an institution. It would, therefore, seem preferable for the contributing countries to take on the continuing burden of paying charges on those SDR which they will have contributed to an institution through a link mechanism. The operation thus becomes similar to the contributing country having borrowed money which is then transferred without interest or charges to the anti-desertification institution. If that institution chooses to hold the contributed SDR, it earns a net return. If it chooses to use the SDR, it can do so without incurring an interest cost.

It is the expectation that an anti-desertification institution receiving SDR via a link mechanism will be applying them for interest subsidy purposes to generate some multiple amount of commercial borrowings under a set of guarantees so that it will, in effect, be using its SDR. However, in so far as it has some flexibility in using them, the form of indirect link where SDR as such are contributed carries with it the prospect of an anti-desertification institution earning a net return. Rather than resort to this form of transfer of SDR allocations, some developed countries may resort to the transfer of an equivalent amount in their local currency. Their choice would be guided by a comparision of the cost of each one of the two types of allocation.

Assuming that the IMF obtains support by October 1981 for the proposal of an allocation for the fourth basic period 1982-86, and while an annual allocation of 10 billion SDR, as proposed by the IMF staff and endorsed by the Group of 24, appears most plausible, the experts who prepared this study chose to limit themselves to a more conservative estimate of 5 billion SDR per annum. If 25 per cent of SDR allocations are thus transferred through the indirect link[14] by countries having a per capita income of at least $3,000 and not in a state of 'structural imbalance' as defined by the Fund, the same amounts involved would reach 800 million SDR.

Supposing 10 per cent of this sum was released for desertification control, 80 million SDR would thus be made available for interest subsidies for such programmes.

D The Common Fund for commodities

Background

The Common Fund (CF) is composed of two separate accounts. The First Account would help to finance the buffer stocks of (existing and new) International Commodity Agreements (ICA). Its pooling of the resources flowing to it in consequence of the association of various ICA would help to achieve economies in financing requirements since the financial needs of different commodities would not generally be synchronous, so that proceeds from sales of commodities with rising prices could often finance purchase of commodities with falling prices.

First Account

The ICA associating with the CF would be required to make cash deposits equal to one-third of the Maximum Financial Requirements (MFR) anticipated for acquisition of buffer stocks. The member governments of the ICA would be required to make guarantees to back the maximum expected borrowing equal to the other two-thirds of the MFR. Together these deposits and borrowings would add significantly to the resources of which the Fund would dispose. Current estimates suggest that, if suitable agreements were reached on the major commodities under discussion (rubber, tin, wood, sugar, coffee, tea, cotton, copper, jute and hard fibre), the total resources of the First Account would be between $6 and $8 billion.

The First Account is, however, not intended as a source of international revenue, nor likely to be authorized to act as such. Its capital stock is not an annual flow but a fixed resource, and it is not expected to be replenished (nor need to be) on a regular basis.

Second Account

Under the agreement constituting the Common Fund,[15] the Second Window programmes include research, marketing, development of commodity processing, and productivity improvements.[16] This Second Account is to have resources of $350 million, consisting of $70 million allocated from the capital of the Fund and $280 million in separate voluntary contributions. The Second Account is subject to replenishment. It would appear, then, that the annual flows of concessional lending through the Second Account would be approximately $100 million in the first three years; subsequent flows would depend upon the volume and frequency of replenishment as decided by the Governing Council of the Fund.

For certain commodities, programmes of development measures may include environmental aspects, subject to the termination by producers and consumers concerned within the International Commodity Body (ICB) designated for such specific commodity by the CF's Executive Board according to paragraph 9 of article 7 of the Agreement. To qualify for such designation, an ICB must meet the following eligibility criteria:

(a) An ICB shall be established on an intergovernmental basis, with membership open to all States Members of the United Nations or of

any of its specialized agencies or of the International Atomic Energy Agency.
(b) It shall be concerned on a continuing basis with the trade, production and consumption aspects of the commodity in question.
(c) Its membership shall comprise producers and consumers, which shall represent an adequate share of exports and of imports of the commodity concerned.
(d) It shall have an effective decision-making process that reflects the interests of its participants.
(e) It shall be in a position to adopt a suitable method for ensuring the proper discharge of any technical or other responsibilities arising from its association with the activities of the Second Account.

Sponsorship by the relevant ICB is a basic requisite for any decision to include desertification-control measures among those eligible for Second Account support for any given commodity. ICBs may eventually be recognized for some 18 or more commodities, each produced by an array of countries that may or may not be affected by desertification. Many – though not all – of these producers may be developing countries (and many of the consumers of these commodities are developed countries). By way of illustration, Tables 4 to 9 below show the world's top ten producers of wheat, coffee and cotton, with output data for 1977 to 1979 (Tables 4, 5, 6), and estimates for 1980 (Tables 7, 8, 9, 10).

Current status

The Agreement on the Common Fund will enter into force when it has been ratified by 90 countries accounting for two-thirds of total capital subscriptions, and 50 per cent of the targeted voluntary contributions to the Second Account. By 20 July 1981, 41 countries had signed the

Table 4

Market rate of interest on borrowings	Leverage factors[a]	
	For disbursement in year	For disbursement spread equally through 7 years
15 per cent	1.5	2.5
12 per cent	1.8	2.8
10 per cent	2.4	3.2
8 per cent	2.8	4.0

[a] These are factors applicable to lending on Third Window Terms with 4 per cent interest and seven years grace period, 25 years maturity. The leverage factors represent the multiples of lending permitted by each dollar of interest subsidy for loans raised at the specified market rates and lent on specified disbursement schedules provided the borrowings are supported by equivalent guarantees collateral.

Table 5 *Top Ten Producers of Wheat (1979)* (in thousands of metric tonnes)

Producer	1977	1978	1979
Union of Soviet Socialist Republics[a]	92,165	120,824	90,100
China[a]	45,001[b]	52,002[b]	60,003[b]
United States of America[a]	55,420	48,922	58,289
India[a]	29,010	31,749	34,982
France	17,350	20,970	19,393
Canada	19,862	21,146	17,746
Turkey[a]	16,720	16,769	17,631
Australia[a]	9,370	18,250	16,100
Pakistan[a]	9,144	8,367	9,944
Italy	6,347	9,332	9,140

[a] Countries affected or threatened by desertification.
[b] FAO estimate.
Source: *FAO Production Yearbook*.

Agreement (although only 8 had ratified it). The signatories account for little more than 61 per cent of the direct contributed capital of the Fund. Pledges to the Second Account were $225 million, out of a target of $280 million. It is noteworthy that OPEC countries accounted for one-fifth of contributions to the Second Account,[17] and non-OPEC developing countries accounted for 15 per cent.

The negotiating conference set a period of 18 months from October 1980 as the period allowed for ratification, with provisions for two extensions, if required. The progress just described suggests that the CF could be established on schedule. However, there could be delays in

Table 6 *Top Ten Producers of Coffee (1979)* (in thousands of metric tonnes)

Producer	1977	1978	1979
Brazil[a]	975	1,226	1,295
Columbia[a]	571[c]	669[c]	762[c]
Ivory Coast	291	196	275[c]
Indonesia	198	223	267
Mexico[a]	182	215	228
Ethiopia[a]	191[c]	191[c]	194[c]
El Salvador	143	132	180[c]
Guatemala	148	149	169[c]
Uganda[a]	150[c]	121[c]	120
India[a]	102	125	105

[a] Countries affected or threatened by desertification.
[c] Unofficial figure.
Source: *FAO Production Yearbook*.

Table 7 *Top Ten Producers of Cotton (1979)* (in thousands of metric tonnes)

Producer	1977	1978	1979
United States of America[a]	3,133	2,364	3,163
Union of Soviet Socialist Republics[a]	2,697	2,731	2,821[c]
China[a]	2,049[b]	2,167[b]	2,207[b]
India[a]	1,208	1,279	1,220[b]
Pakistan[a]	553	475	650
Brazil[a]	555	477	575
Turkey[a]	575	475	505[c]
Egypt[a]	409	438	482
Mexico[a]	418	340	336
Guatemala	137	147	146[c]

[a] Countries affected or threatened by desertification.
[b] FAO estimate.
[c] Unofficial figure.
Source: *FAO Production Yearbook*.

view of the situation of the ICA. At least one major industrial country intends to withhold ratification until a sufficient number of 'suitable' ICA are available to associate themselves with the CF. At the present time, it appears that the International Cocoa Agreement, the International Rubber Agreement and the International Tin Agreement meet the criteria for consistency with the CF, in that they rely primarily on buffer stocks. The international agreement on coffee relies primarily on export quotas instead of buffer stocks and would not be considered eligible for association on the CF in its present form. The Sugar Agreement is based on internationally co-ordinated national stocks which are also eligible for financing by the CF. However, the agreement

Table 8 *Wheat Production, estimated 1980* (in millions of metric tonnes)

Major Producers	
Union of Soviet Socialist Republics[a]	98
United States of America[a,b]	65
China[a]	57
European Community[b]	51
India[a]	32
Canada[b]	19
Turkey[a]	17
Pakistan[a]	11
Australia[a,b]	11
Argentina[a,b]	8

[a] Countries affected or threatened by desertification.
[b] Five major exporters.
Source: International Wheat Council, London.

Table 9 *Coffee Production* (Exportable production estimated for crop year October 1980–September 1981) (in thousands of 60 kg bags)

Major producers/exporters	
Brazil[a]	16,350
Colombia[a]	10,575
Indonesia	3,918
Ivory Coast	3,117
Mexico[a]	2,600
El Salvador	2,350
Guatemala	2,223
Uganda[a]	2,134
Ethiopia[a]	1,700
United Republic of Cameroon[a]	1,662
Costa Rica	1,524
Ecuador[a]	1,496
India[a]	1,333
Kenya[a]	1,297
Honduras	1,280
Zaire	1,205
Peru[a]	1,100
Madagascar[a]	1,062

[a] Countries affected or threatened by desertification.
Source: International Coffee Organization, London.

Table 10 *Cotton production, estimated 1 August 1980–31 July 1981* (in thousands of metric tonnes)

Major producers	
United Soviet Socialist Republics[a]	3,079
China[a]	2,624
United States of America[a]	2,428
India[a]	1,336
Pakistan[a]	672
Brazil[a]	568
Egypt[a]	528
Turkey[a]	484
Mexico[a]	347
Guatemala	136
Argentina[a]	130
Syrian Arab Republic[a]	119
Greece	116
Colombia[a]	112
Peru[a]	103
Sudan[a]	101

[a] Countries affected or threatened by desertification.
Source: International Cotton Advisory Committee.

does not provide for international financing of the acquisition costs, which would be essential if it is to associate with the CF. For copper, there appears little possibility at the present time that an ICA will be established, although considerable discussion has taken place on an arrangement based on an international buffer stock.

Financial implications

Even over the medium and longer term, when it seems likely that the CF will be ratified and operational, it is unlikely to be a major source of incremental revenue for general development. For its first three years, the annual concessional flows of the Second Account will amount to only approximately $100 million, or 0.3 per cent of global annual concessional assistance, although this amount could grow, depending on future replenishments of the Second Account. Even this extremely modest flow may not necessarily be 'additional' to other aid flows, in that donor countries may merely rearrange the composition of their normal foreign assistance programmes to make funds available for the CF without increasing total aid.

The Common Fund for Commodities, once properly constituted, can make a significant contribution to economic growth and stability for both North and South through commodity price stabilization. Furthermore, it is unique among international financial institutions as a creation primarily of the developing countries (rather than a product of the Bretton Woods era), which will have a much larger role in voting and control than in most other international financial institutions. But this does not make the Common Fund a major source for additional resources for international development in the more narrowly-defined sense of extra tax revenue or source of concessions funds for international purposes, such as desertification.

E International revenues for sea-bed mining

Background

The seas have long been considered one of the primary 'international commons'. They have thus been indentified as one of the most likely potential resources to be used for the benefit of the international community. Long before the Brandt Commission's reference to such use,[18] the Declaration on Principles Governing the Sea-Bed and the Ocean Floor, and the Subsoil thereof, beyond the Limits of National Jurisdiction (General Assembly resolution 2749 (XXV) of 17 December 1970) stated that the sea-bed and its resources formed part of 'the common heritage of mankind' and laid a foundation for the subsequent negotiations in the Third United Nations Law of the Sea Conference toward the establishment of an international legal regime dealing with the uses of the sea. This regime would include, *inter alia*, the rights of passage for civil and military shipping, and a vast range of definitions of other legal and economic aspects and issues, including safeguarding the interest of the developing countries. Section 4 of Part XI of the United Nations draft Convention of the Law of the Sea ('The Convention') addresses the broad questions of activities in the area, including the development of both living and non-living resources. The International Sea-Bed Authority ('The Authority'), which is to be composed of all United Nations Member States, would govern operations relating to sea-bed mining and would be the sole body authorized to levy taxes, fees, royalties, or other types of charges on such operations.

The draft Convention would establish an exclusive economic zone within 200 nautical miles of a nation's coastline (article 57). In addition, the draft Convention would grant nations exclusive economic rights over the continental shelf (the 'submerged prolongation of the land mass of the coastal State') (article 76 and 77). The draft Convention would declare the sea-bed beyond national jurisdiction to be the common heritage of mankind and places jurisdiction over the sea-bed in an International Sea-Bed Authority (articles 136, 137). The Authority, acting through its various organs, would be required to approve all sea-bed mining projects, including those of private or state-owned commercial operators or of consortia composed of two or more of these, or those in which it chooses to engage itself through its operating entity ('The Enterprise'), acting by itself or as a joint venture partner. Accordingly, the only channel for tapping the large potential that such activities may generate in the future would lead through the Authority. There is no present alternative legal basis for levying either specific national or international taxes on such activity, regardless of its purpose; nor do the provisions of the draft Convention lend substance to any suggestion of there being a viable economic rationale for any such alternative method of taxing the economic rents accruing to any or all of the different parties that may, in due course, decide to engage in such activity, regardless of whether such parties are private or public entities operating outside the Authority, or the Authority's own operating arm, referred to hereinafter as 'The Enterprise'.

The 1958 Convention on the Continental Shelf[19] had already granted national jurisdiction in coastal waters up to 200 metres in depth, and in deeper waters 'adjacent to the coast'. The present draft Convention moved toward the establishment of exclusive national zones up to 200 nautical miles offshore and, in certain cases, beyond. Actually, the draft Convention would provide for certain payments in cases of exploitation of non-living resources where the continental shelf extends beyond the

200 mile limit. One main significance of this consensus toward the 200 mile limit in the United Nations Convention on the Law of the Sea is that it sharply reduces the potential for internationalizing some or even all of the oil and gas reserves lying beyond. The North Atlantic, for instance, includes the relatively shallow Grand Banks, which are widely believed to hold such potential, even though its actual exploitation still faces formidable technical obstacles in an area hundreds of miles from dry land and exposed to the risk of weather and iceberg movement.

With petroleum reserves thus essentially ruled out as a source of international revenue, the mining of manganese nodules from the deep sea-bed beyond national jurisdiction remains the only presently evident potential source for such revenue. These potato-sized nodules, which are strewn on the floor of the deep sea, average around 30 per cent manganese by weight; but their economic value derives from nickel, followed by copper, manganese and cobalt, and possibly molybdenum and vanadium, contained in them. Several mining and metallurgical technologies have been developed, proven in the laboratory and, to a degree, on experimental and/or test sites. They result in the production of either three metals (nickel, copper and cobalt), or four (including manganese), or five (including molybdenum) as joint products in relatively fixed proportions. But no full-fledged pilot plant has ever operated for the requisite period of time, which would be the prime precondition for scaling up to a full-sized operation. In fact, while individual operators have probed further into the technical problem, official authorities raised new demands for costly prototype demonstration units. Together with subsequent changes in metal prices and production costs, these developments may go some way toward explaining why such operators seem to have downgraded the commercial priority accorded to sea-bed as against land-based mining projects since the time when earlier projections had raised high hopes for their revenue potential. This downgrading has, *inter alia*, resulted in a postponement of the most likely start-up schedule, from an earlier expectation of a first venture period between 1982 and 1985 to one falling more toward the early part of the following decade, with the current outlook pointing toward 1992. Consequently, another element of uncertainty has been added to the hazards of estimating tonnage production volume, prices and costs of deep sea-bed mining.

In principle, the draft Convention recognizes three sources of potential revenue from mining beyond the 200 mile limit:

(a) Taxes ('shares of net proceeds') and royalties ('production charges') from exploitation of non-living sea-bed resources;
(b) Income from operations by the Enterprise;
(c) Payments and contributions in kind with respect to the exploitation

of the continental shelf beyond 200 miles as referred to earlier (developing States which are net importers of a mineral resource produced from their continental shelf are exempted from payments with respect to that mineral resource).

These three types of income accruing to the Authority would be distributed in accordance with articles 173 of the draft Convention[20] which directs that funds remaining after payment of the Authority's administrative expense may, *inter alia*, go to States that are parties to the Convention on the basis of equitable sharing criteria; the draft Convention defines these criteria in article 140 with respect to sources (a) and (b) above, while it deals with the type (c) income, which is of relatively smaller proportions, in article 82. The difference between these two articles arises from the fact that article 140 covers revenues arising from operations lying within the area – the Authority's main zone of authority, i.e. beyond national jurisdiction – whereas article 82 addresses operations from extensions of the continental shelf, which remain outside the area as defined by the draft Convention. This distinction is crucial when it comes to the distribution of any income accruing *to* the Authority itself from operations within that area. The relevant provision of article 140 specifies that such operations be carried out for the benefit of mankind as a whole. Under paragraph 2 of that article, the Authority, in effect acting as a principal, 'shall provide for the equitable sharing of financial and other economic benefits deriving from activities in the area through any appropriate mechanism, on a non-discriminatory basis'. The language of article 140 thus remains somewhat less specific than that contained in article 82, under which distribution of the payments or benefits from continental shelf operations through the Authority (i.e. as agent) would be 'on the basis of equitable sharing criteria, taking into account the interests and needs of developing States, particularly the least developed and the land-locked amongst them'. It is this particular provision of article 82 that presents a legal basis for advancing the interests of a group of countries that includes those most adversely affected by such environmental threats as those posed by the spread of desertification. Given the somewhat greater leeway for defining the criteria for distributing income from the Authority's own activities, which provide the bulk of prospective revenues under headings (a) and (b) above, the group of experts would urge the Authority, once constituted, to adopt criteria essentially similar to those defined in article 82 in distributing income from these latter sources. This would seem to be the most effective way of raising additional revenue for the broad purposes contemplated by the terms of reference proposed by the group. As regards the predictability of such future revenues, the economic specifics of sea-bed mining, starting with the prospective impact of the production ceilings referred to above, are discussed below.

Past analyses of the economics of sea-bed mining have indicated that the supplies of cobalt and nickel could exert significant adverse effects on existing producers of some of the nodule metals. Earlier studies concluded that the contribution of manganese nodules to world production of copper and nickel would not be sufficient to affect prices to any significant extent, but that it would create sufficiently large additions to the production of cobalt that could substantially reduce its price.[21] This is also corroborated by the findings of later studies carried out for the Committee on Natural Resources of the Economic and Social Council, 'Mineral Resources: Trends and Salient Issues, with particular reference to molybdenum, cobalt and vanadium, including problems of Technology Transfer' (E/C.7/115, 6 April 1981). Section F of chapter II of that study, after describing the status of negotiations at the Third United Nations Conference on the Law of the Sea, presents an analysis of the effect of the production ceilings which, under the draft Convention formula, would be calculated from the trend line of nickel consumption. In its basic form, the draft Convention's formula would provide that the sea-bed production of nickel may not exceed (a) an amount equal to the entire increment in trend-line world consumption of nickel during the five-year period immediately preceding sea-bed production (which now seems likely to begin by about 1992), plus (b) an amount equal to 60 per cent of the trend-line increment in world consumption thereafter (article 151). The trend line underlying the data shown here, by way of illustration, projects historical data derived from consumption over the last 15 years to the years 1990 and 2000.

Shown in Table 11 below are ranges of production ceilings for four

Table 11

	1990		2000	
	Low[a]	High[b]	Low[a]	High[b]
Nickel	196,000 tons (21)	252,600 tons (20)	320,700 tons (28)	806,500 tons (39)
Copper	168,600 tons (1.0)	218,000 tons (1.4)	275,800 tons (1.2)	693,600 tons (2.9)
Cobalt	21,600 tons (58)	27,900 tons (74)	35,300 tons (89)	88,700 tons (169)
Manganese	1,176,000 tons (8)	1,521,600 tons (10)	1,924,200 tons (10)	4,839,000 tons (25)

[a] Low case assumes that world nickel demand increases at 2 per cent beyond 1979.
[b] High case assumes that world nickel demand increases at 5 per cent beyond 1979.
NB: Figures in parentheses are percentages.
Source: E/C.7/115, p. 22, para. 57.

metals based on a study entitled 'Effects of the production limitation formula under certain specified assumptions' (A/CONF.62/L.66). The figures shown can do no more than illustrate orders of magnitude of the maximum amounts allowable from sea-bed mining. They are predicated on such operations starting in 1988.

The figures in parentheses indicate the ratios of the production ceilings to the hypothetical world demand figures for the respective metals. In the case of nickel, the high case, these data indicate that sea-bed supply could reach about one-fifth of world production in the third year of production, and nearly 40 per cent by the end of the century.

Clearly, the production limitations in the draft Convention reflect compromises between the interests of existing producers, as opposed to those engaging in sea-bed mining entities and, in a wider sense, the international community at large. As indicated, some existing producers feel that this present compromise does not reflect an adequate balance, and therefore they oppose that aspect of the draft Convention because it would not provide for any compensation for the potential losses of independent operators (article 173 only protects States), as part of the distribution of the Authority's net income. However, earlier studies[22] indicate that such losses could assume major dimensions relative to the prospective economic rents from sea-bed mining, thus effectively negating the draft Convention's revenue-raising objectives. It remains to be seen how this further, and major, source of uncertainty will be dealt with in the course of future negotiation.

Another central feature of the draft Convention is that it would give the Authority direct access to mining operations through the Enterprise, which could do so either by itself or through joint ventures with independent operators. Any independent operator – in practice, mainly consortia organized for the purpose – applying to the Authority for a mining license would be required to propose two mining sites, one of which the Enterprise could choose for itself. In that case, the Enterprise would be entitled to a certain proportion of the production ceiling. In addition to this parallel exploitation of sea-bed resources, the draft Convention would also require independent operators to sell their technology to the Enterprise at a fair market price. The Convention's attempt to find common ground between reliance on the private market, and the internationalization of production advocated by some developing States, have caused a degree of unease among independent operators. Obviously, without resolution of these issues it is difficult to offer more than the most tentative of projections of the revenues flowing from the Authority's operations.

In addition to the profits earned by the Enterprise, international revenue would be obtained from income taxes and production charges from the actual mining operation, but not from the related transportation and processing activities. It is assumed that these would remain subject to national taxation, although the draft Convention does not address this point. To the sea-bed miner, the draft Convention would actually offer a choice between a so-called single system and a combined, or mixed, system. The single system would suit collective economies which seldom resort to taxing incomes; instead, it would only call for a charge on production, ranging from 5 per cent in the first ten-year period of production to 12 per cent in the second period, which starts in the eleventh year of commercial production. These rates would yield revenues roughly in harmony with those obtained under the so-called mixed system, which involves levying a combination of production charges and actual taxes. That mixed system would impose:

(a) Up to the time of full recovery of investment (with interest), a 2 per cent flat charge on gross sales plus an incremental income tax on the profits from the mining sector of 35 per cent, 42.5 per cent, or 50 per cent, depending on the rate of return on investment achieved – up to 10 per cent, 10 per cent to 20 per cent, and above 20 per cent.
(b) Thereafter, the flat charge doubles to 4 per cent on gross sales, plus profits taxes of 40 per cent, 50 per cent, or 70 per cent, again depending on the same scale of return on investment.

In addition to these international revenues accruing to the Authority, the typical integrated operator would be liable to national income taxation in the jurisdiction governing his related activities in transportation and processing. This raises the problem of transfer pricing, i.e. the price which the mining operation charges the downstream operations described, because that price would determine the ratio in which taxable incomes are allocated as between international and domestic authorities. This was another point of contention during the negotiations; it was finally resolved by reference to the United Nations Commission on Transnational Corporations and the Expert Group on Tax Treaties between developed and developing countries. These bodies are to establish criteria for fair transfer pricing by reference to free market or arm's length transactions in relevant markets.

The issue of national authorities granting relief, in the form of full tax credit or else allowability as a tax deduction, while not addressed in the draft Convention, can be answered analogously to the same problem in the proposed taxation of communications satellites (in the following section of this chapter). For the reasons stated there, no country allows tax credits against international levies, leaving deductibility as the only protection against what might turn out to be unacceptably high levels of over-all taxation for those entities engaging in sea-bed mining that are liable for domestic taxation of their downstream operations. Accordingly, industry pressure on national legislatures for more extensive relief than currently available would appear likely. Conceivably, the extent to which such pressures are accommodated could, in turn, influence the pertinent international revenues, given their dependance on rates of return.

Possible magnitudes of international revenue

Geological, technological and economic uncertainties, along with the legal issues that remain unresolved, render any projection of possible magnitudes of international revenue an extremely difficult task. Information about crucial geological factors, e.g. abundance and metal content of manganese nodules, topography of the area to be mined, is sparse. While the basic technology for mining, transporting and processing manganese nodules exists, it has to be further tested for efficiency and reliability. The workability, over a period of time, of the total system, integrating all the components, has to be demonstrated and finally the total system has to be scaled up to commercial level. Metal markets are notoriously volatile and projections of future demand, supply and price levels vary widely. Lastly, as indicated earlier, the provisions of the legal regime, as reflected in the draft Convention on the Law of the Sea, are negotiating provisions, so that the final nature of the provisions still remains an open question.

A number of studies have examined the economic dimensions and profitability of sea-bed mining. The wide divergence of the conclusions can be seen from the estimates of internal rates of return of a sea-bed mining project ranging from as low as 5 per cent to as high as 100 per cent.[23] Virtually all these researchers and industry experts agree that considerable uncertainty remains about the profitability of sea-bed mining, depending on the geological, technological, economic and legal factors.

The most recent and detailed estimate of the profitability of a sea-bed mining project are to be found in the studies by the Massachusetts Institute of Technology (MIT) and Arthur D. Little. The MIT study constructs a computerized model which allows introductions of variations centered around a baseline case. On the basis of certain assumptions about, *inter alia*, abundance and metal content of manganese nodules, efficiency of mining and processing technology, capital and operating costs, future prices of metals and capital structure of the project using a 1:1 debt equity ratio, the baseline case shows an internal rate of return of 18 per cent. Taking the MIT baseline case, and several variations around that case, and applying the taxation provisions of the draft Convention, the Chairman of Negotiating Group 2 of the Third

United Nations Conference on the Law of the Sea (the group which dealt with the financial and fiscal provisions of the draft Convention) estimated the revenues of the Authority over a 25-year lifespan of one sea-bed mining project to be in the range of $260 million to $1,960 million in 1976 US dollars[24] with the following additional assumptions:

(a) commercial sea-bed mining would start in 1992;
(b) five sea-bed mining projects would start in that year;
(c) the Authority's income would be distributed evenly over 25 years;

the international revenue from sea-bed mining in 1992 could be estimated to range from about $50 million to $390 million. This range of estimates thus falls below the Brandt Commission's earlier estimate of $500 million by the middle or late 1980s.[25]

Assuming 3 per cent annual growth in world consumption of nickel, the production limitation formula discussed above would allow roughly 10 sea-bed mining projects by the year 2000, and international revenue would likewise double by the year 2000.

When or whether the costly next steps in sea-bed mining will be taken will largely depend on the future development of the legal regime and on the future outlook in the world metal markets.[26]

It may be noted that United States legislation in effect since 1980 permits firms to register claims, provides the basis for exploration, and permits them to begin actual mining if international agreement is not reached by 1988. Similar legislation exists in the Federal Republic of Germany. The United States law also provides for a trust fund for taxes on sea-bed mining (although at considerably lower rates than the draft Convention) for possible disbursement to an international entity if an international agreement is reached.[27]

Despite the possibility of national legislation, the mining consortia may be unprepared to proceed with investments without an international treaty. The investments are large, as much as $1 billion per operation, and in the face of uncertain claims subject to international reversal, firms are unlikely to risk such sums, and capital markets are unlikely to lend them.

F   Taxes on 'parking fees' from geostationary communications satellites

Background

A geostationary earth satellite rotates in a unique orbital zone around the equator, at a distance of approximately 22,300 miles from the earth, travelling from west to east, and remaining in approximately the same position above the earth at all times. These satellites have become central to modern communications by transmitting signals from telephone, telegraph, teletype and facsimile systems for both domestic and international commercial, military, meteorological, maritime, and numerous other uses, some of which serve environmental purposes (A/AC.105/203 and Add. 1). There is a limit to the number of such satellites that the geostationary technologies orbit zone can accommodate. Satellites must be spaced at certain intervals to avoid interference among their communication signals.

The International Telecommunication Union (ITU) determines which parts of the frequency spectrum are available for satellite communications. At the 1979 World Administrative Radio Conference ITU established procedures for the assignment of frequencies by its International Frequency Registration Board (IFRB) so as to avoid early saturation of the frequency spectrum. There is, however, no international system for allocating positions in the geostationary orbit zone. The thought of an eventual scarcity of such positions, sometimes referred to as 'parking slots', has suggested to some observers the concept of imposing fees for their uses,[28] or else taxing revenues on profits from such use. The Brandt Commission cited satellites as a possible new source of international revenue to be used for economic development,[29] a suggestion already made in the 1980 study on financing the Plan of Action to Combat Desertification (Part II of this book).

The developing countries' present lack of direct access to the requisite technologies does not bar them from sharing the benefit of geostationary satellites, since all of these countries are free to use the facilities of INTELSAT, and many of them do. Meanwhile, their access rights to radio frequencies in their area enjoy the protection provided for the 1966 Treaty on Principles Governing the Activities of States in the Exploration and Use of Outer Space, including the Moon and other Celestial Bodies (General Assembly resolution 2222 (XXI), annex) to the degree that the IFRB allots frequencies within three regions, each of which includes both industrial and developing countries. This, to a degree, mitigates the more advanced countries' inherent advantage from applying a 'first come, first served' satellite system. It also suggests the possibility of certain developing countries leasing out, for finite periods of time, their frequencies to countries in need of additional frequencies for their communication satellites. While this would provide the developing countries involved with additional, though temporary, revenue, there would be no confluence of such funds for common developmental goals.

Prospective demand growth could complicate the task of coordinating the use of communications satellites. However, technological progress may well determine the degree to which expectations of scarcity will prove to be realistic. The ITU is convening another World

Administrative Radio Conference which will discuss various communications problems in two sessions in March 1984 and in November 1985, and these meetings could consider the imposition of taxes or fees for assignment of satellite parking slots. [No conclusive decisions were reached at these sessions.]

Taxation principles

The 1966 Treaty on the use of outer space provides that, although exploration and use of outer space 'shall be the province of all mankind' (art. I), national jurisdiction remains binding because a 'State Party . . . on whose registry an object launched into outer space is carried shall retain jurisdiction and control over such object' and 'ownership of objects launched in outer space . . . is not affected by their presence in outer space' (art. VIII). National taxation, therefore, is the norm, and not international taxation.

International taxation of satellites would, therefore, appear to be ruled out, but the possibility of international taxes or charges for the use of geostationary parking slots is left open. Such taxes or charges should not, however, exceed the rental value of the slots, or they would prevent the owners of satellites from earning normal profits and thereby discourage satellite activity.

The slots have rental value only if they are now, or are expected within a reasonable time to be, a scarce resource. Slots well-located with respect to important land areas already are scarce resources and a situation of general scarcity of slots could arise before the end of the century.

Geostationary orbital positions are analogous to sea-bed resources that are scarce, finite, and may be said to belong to 'all mankind'. Therefore, international collection of the rental value of their use is justifiable. It is difficult to decide how much the rental value is, and therefore how high the tax should be, if the tax is to be set arbitrarily by an international forum, such as ITU. A more efficient approach would be to allow competitive bidding for satellite parking slots. The revenues from these bids could then be made available for international uses, in the form of annual fees for the duration of the slot assignments. Many national telecommunications systems use taxation or competitive bidding to allocate the available radio and television frequencies domestically, and the same principle could apply internationally to geostationary orbital positions.

Competitive bidding for parking slots would be a more efficient system of preventing overcrowding than arbitrary rationing. Taxation, even without competitive bidding, would also be better than rationing, because it would limit the number of satellites by favouring those that would make the best use of particular slots.

If international taxes are imposed on the use of geostationary orbital positions, the question of double taxation would arise. The situation would be similar to the taxation of a firm's income by two nations. The established solution in such cases is for each nation to permit the firm to deduct the tax paid to the other nation from its taxable income. (This arrangement, it might be noted, is not the same as a tax credit, or full deduction of levies paid another nation from tax liability.) It would probably be necessary to obtain agreement from the countries concerned that international taxes would be deductible from income subject to national taxes. Such an arrangement would not be based on reciprocity, as is the case in similar arrangements between nations, but on equity and the need to avoid excessive taxation of satellite activities.

Prospects of overcrowding

At the present time, there are approximately 100 active geostationary satellites in orbit, out of a total of 126 geostationary satellites placed in orbit since 1963. Nearly half of that total, however, is dedicated to uses other than those related to civilian communications, such as military, maritime, meteorology, space research and remote sensing – all activities that limit the applicability of income-based taxation, if not that of fees for parking or royalties based on utilization. Current technology requires that geostationary satellites be spaced at least three degrees apart. Under the technology of standard satellites of the past, the most heavily used portions of the orbit (over North America, the Indian Ocean, and the Atlantic Ocean) already have satellites spaced as closely as possible.

Technology is changing rapidly, however. Recent satellites placed into orbit have achieved greater traffic volume by frequency re-use through polarization, whereby some signals are sent vertically and others horizontally, and by the use of directional antennas permitting more than one satellite in the same orbital position to serve different areas. New satellites are using the separation of beams to permit separate transmission on each beam.[30]

Technology is expected to change further in the medium-term future. The spacing of satellites may be narrowed to less than three degrees. Frequencies may be used that are not currently used. Satellites may be 'stacked', one at 1.5 degrees to the north of the equator and another at 1.5 degrees to the south, to permit dual use of a single orbital slot without interference. Within a decade as many as 10 to 12 satellites might be strung together in a halo or figure eight configuration at a single orbital slot, multiplying capacity tenfold. In the longer-term future, technology may move towards 'antennae farms', or large space platforms designed for multiple use, permitting fewer satellites with much larger capacity to handle telecommunications and other services. Some observers doubt that foreseeable demand growth will be strong enough to permit any real scarcities to develop assuming full and co-ordinated exploitation of the technological innovations currently in prospect.

However, the demand for satellite services is also growing rapidly. INTELSAT, which operates 12 of the geostationary satellites, has had a growth in the number of telephone circuits from 1.8 thousand in 1969–1970 to 14.4 thousand in 1978–1979, an annual average growth rate of 26 per cent.[31] At this rate demand doubles in three years and multiplies tenfold in a decade. In other words, technological changes that could multiply the capacity of a geostationary satellite tenfold by 1990 could well be needed just to keep up with the pace of increasing demand.

According to a recent United Nations analysis entitled 'Efficient Use of the Geostationary Orbit' (A/CONF.101/BP/7), the best estimate at the present time is that foreseeable technological improvements will make it possible to meet the growing requirements for geostationary satellite service for at least the next two decades without encountering a generalized problems of overcrowding. For the foreseeable future, the instances of scarcity seem likely to be limited to the choicest locations.

Financial considerations

Even though there appears to be little prospect of general overcrowding of orbital positions for geostationary satellites for the next two decades or so, it is informative to consider the financial magnitudes that might one day be involved in international taxation of these satellites. The annual revenues of INTELSAT provide a basis for gauging the size of the market for services of geostationary satellites. In 1980 INTELSAT had total revenues of $215 million.[32] Its depreciation and operating expenses accounted for $112 million, leaving (untaxed) profits of approximately $100 million. Considering that INTELSAT accounts for between one-third and one-fourth of the geostationary satellites in operation, a conservative but purely illustrative estimate might be that the total market value of services of such satellites was $500 million in 1980. This allows for the fact that some satellites produce no revenue. While many of the other United States satellites are commercial (for example, the RCA, Western Union, and COMSAT satellites used for domestic United States telecommunications), a portion is used by defence and other public entities.

Assuming that profits are approximately 50 per cent of revenue (as for INTELSAT), an international tax of 50 per cent imposed on profits would generate $125 million annually at the present time; alternatively, a 10 per cent royalty on total revenue would generate $50 million, and an annual 'parking fee' of $1 million per satellite about the same amount. Compared with total concessional economic assistance of approximately $30 billion in 1980,[33] these figures are not large; if 10 per cent were to go toward desertification projects, the result must be viewed as minimal relative to the $1.8 billion annual target.

In view of the exceedingly high growth of demand for telecommunication satellite services, however, these financial dimensions could become much larger in 10 or 20 years. Even assuming a more modest growth rate of demand than in the past – for example, 15 per cent annually – total revenue from all geostationary satellites would grow from a postulated $500 million in 1980 to $2 billion in 1990 and $8 billion in the year 2000 (at 1980 prices). The need to resort to greater technological sophistication would probably reduce the ratio of profits to revenue, but under optimistic assumptions that perhaps 40 per cent of revenue might be profit, a 50 per cent international tax on profits would generate $400 million by 1990 and $1.6 billion by the year 2000 (at 1980 prices).

A 50 per cent tax rate would require a full tax credit by national governments, meaning, in effect, a donation on their part of the entire regular income tax on the private operator. In fact, most industrial countries' effective tax rates lie below that figure; thus, full credit for a 50 per cent tax would require such governments to give up more than the full tax, resulting in either a reduction of the rate at which non-satellite operations are taxed, or else a net credit to be refunded to the taxpayer. It seem extremely unlikely that this would find acceptance among national governments. If, instead, the 50 per cent 'international tax' were to be treated as a tax-deductible expense, the result would be an effective halving of the operator's after-tax profit – clearly an extremely high burden. The alternative approach of using a royalty of, say, 5 per cent to 10 per cent on gross revenues would be more acceptable to both taxpayers and governments. It would also offer greater predictability, in that the yield would fluctuate along with revenue which is less volatile than operating costs, interest and depreciation. Based on the same assumptions regarding demand growth, one can project royalties at rates of between 5 per cent and 10 per cent to generate the following income streams over the next 20 years:

Year	Annual Revenues	At royalty rates of 5 per cent	10 per cent
1980	$500 million	$25 million	$50 million
1990	$2,000 million	$100 million	$200 million
2000	$8,000 million	$400 million	$800 million

Annual parking fees of between $0.5 million and $1.0 million for each taxed communications satellite, rising from 50 in number in 1980 to 200 by 1990 and 800 by the end of this century, would yield exactly the same amount as would the royalties at the rates of 5 per cent and 10 per cent shown above. If use of larger satellites resulted in smaller numbers the parking fees could be raised.

These financial estimates are merely illustrative, but they do suggest two conclusions. First, in the next few years, international revenue from geostationary satellites would probably be small. Second, the expected rapid growth of this market does mean that, by the turn of the century, taxation of these satellites might generate a substantial amount of revenue (equal, in real terms, to as much as roughly half of the current annual lending of the IDA, for example).

Other considerations

The introduction of an international tax on geostationary satellites would encounter institutional obstacles. INTELSAT is already an international entity with over 100 members and over 140 countries use its services. Some would argue that the international community already benefits from this organization, and that imposing taxes on it for international purposes would be redundant or self-defeating. However, by far the largest users of INTELSAT are the industrial countries, so that on balance the developing countries would clearly benefit from taxation with tax revenue earmarked for their economic development, including environmental programmes.

If technology does not manage to keep ahead of the growth in demand for geostationary satellite services, there is a danger that political rationing would allocate the available parking slots instead of market forces through taxation or (better) the competitive auctioning of these slots. At one extreme, rationing could favour the early entrants, the United States and other industrial countries. At another extreme, a rationing system arbitrarily preserving some portions of the slots for developing countries might occur as the consequence of international voting structures, and such an outcome would be unlikely to be as efficient (or, for that matter, equitable, since the developing countries capable of launching satellites would hardly be representative of the low-income countries). Thus, some form of market rationing of these positions through international taxation or competitive bidding would be highly desirable.

Conclusions

For the near term, and very likely much of the next decade or two, international taxation of geostationary satellite parking slots holds little promise as a substantial source of international revenue.

Establishing the principle of international taxation of these unique geophysical resources would, however, be desirable before the possibility is pre-empted by national claims or other arrangements. Moreover, a system of regulating the use of geostationary orbital positions should be established to prevent overcrowding before the need becomes acute. International taxation of orbital positions could be a central feature of such a system. The study recommends that the possibility of early action on these questions be considered by the Preparatory Committee for the Second United Nations Conference on the Exploration and Peaceful Uses of Outer Space, which is to be held in 1982. [The second UN Conference on the Exploration and Peaceful Uses of Outer Space held five meetings between 22nd March and 6th April 1982. The report of the Prepcom contains only two pages of a procedural nature.]

It is important that the international regime governing allocation of geostationary orbital locations evolves in a way that is consistent with international taxation and probably also with competitive bidding.

IV DETAILED MODALITIES OF OBTAINING RESOURCES ON A CONCESSIONARY BASIS

Introduction

The first avenue for raising resources is not to seek grants from governments but rather interest free loans. While current budgetary constraints affecting governments may apply to interest free loans no less than to grants, it is somewhat more plausible to suppose that, since the amounts are repayable, they are more likely to be forthcoming, provided the amount being sought as loans from each government is judged to be reasonble. This route has the advantage that if borrowing countries are charged rates of interest on the order of 2.5 to 3 per cent, a period of no more than 20 years will suffice for the loan to be repaid to donor governments on the basis of accumulated interest alone.

The second avenue consists of raising funds on commercial terms from governments and then utilizing a combination of budgetary grants and other available extra budgetary sources to subsidise the interest rates on lending. This route has typically been resorted to by the IMF.

The third avenue open to governments to raise concessionary resources is to tap private capital markets where obviously loans can only be raised on a commercial basis in terms of both prevailing interest rates and maturities. For the proceeds of such loans to be re-lent on concessionary terms two elements are required:

(a) an interest subsidy element large enough to permit interest rates to the borrower to fall to acceptable levels and to extend maturities sufficiently beyond those of market borrowings to keep the borrower's debt service burden within safe limits;
(b) a system of supporting guarantees and/or collateral that will carry sufficient credibility in the market place to enable the loans to be raised.

This was the procedure adopted in the case of the World Bank's Third Window, where the availability of a subsidy element of a grant character sufficed to transform market borrowings at market maturities, e.g. 8 per cent 10 year bonds, into Third Window terms, i.e. lending to developing countries at 4.5 per cent, with a seven-year grace period, and a 25 year maturity. The guarantee element to support the borrowings on market terms was supplied by the callable capital of the World Bank. In the Third Window as originally proposed,[34] it took $225 million of subsidy to convert a billion dollars worth of market borrowings into lending to developing countries on Third Window terms, a leverage factor of approximately 4 to 1. The advantage of this is simply that relatively modest amounts by way of interest subsidy would suffice to raise a substantial multiple of loanable funds provided that supporting guarantees/collateral satisfactory to the market-place were forthcoming. Again in the case of the World Bank's Third Window, the limiting factors were not only the amounts that could become available as interest subsidy but also the limits then set to the scale of borrowing by the 'unused' callable capital of the Bank before the pending capital increase could come into effect. In general, the multiple of resources that can be raised in this way per dollar of available subsidy depends on the spread between market rates and the desired concessional interest rate, the required degree of maturity extension beyond market terms and the period of disbursement of the loan which, in the case of the Third Window, coincides with the seven-year grace period.

It is this avenue which this section of the report will attempt to explore on the basis of the following assumptions:

(a) The amounts available for interest subsidy purposes are assumed for purposes of illustration, to come from extra budgetary sources (in line with present day realities) and specifically from amounts within the purview of the IMF. These are relative to:
  (i) reflows of money to the IMF arising out of its gold sales operations which were lent through its Trust Fund to developing country members and which returns to the Fund as these loans are repaid;
  (ii) the establishment of a link between SDR and development

finance during the forthcoming fourth five-year basic period for SDR allocation 1982–86.
(b) The subsidy amounts that can reasonably be earmarked for anti-desertification purposes amount to no more than 10 per cent of the total resources that can be released to the international community through the mechanisms outlined in (a) above. In terms of the feasibility studies of (a) (i) and (ii) above and the 10 per cent assumption in (b) above interest subsidy, amounts of 80 million SDR for the years 1982–85 and 105 million SDR for the years 1986–89 would be forthcoming (see Table 2, page 118).
(c) The concurrence of the developed country members of the international community can be obtained to these arrangements on the basis that these need not for present purposes require an amendment of the IMF Articles of Agreements with Executive Board decisions alone providing the necessary authority for their implementation, with appropriate qualified majorities.
(d) An acceptable degree of concessionality is assumed to consist of approximately World Bank Third Window terms, i.e. 4 per cent interest; seven-year grace period; 25 years maturity. Concessioned anti-desertification lending would therefore comprise two separate streams; a highly concessional stream stemming from interest free loans to be provided by governments to an independent institution for lending with maturities up to and over 40 years, and slightly harder Third Window terms permitted by extra budgetary subsidies.
(e) The supporting system of guarantees/collateral would be forthcoming on the following alternative bases to support Third Window terms:
  (i) *ad hoc* arrangements extending during the period 1982 to 1989 to permit the independent corporation for combating desertification or some other entity to raise in capital markets the loans warranted by available subsidy amounts. This would most suitably take the form of a system of 'limited joint and several guarantees' negotiated among interested donors;
  (ii) in lieu of such an *ad hoc* system of guarantees the incorporation in the institution of a callable capital component having the same legal effect and no larger than necessary to permit the nine-year borrowing programme warranted by available subsidies;[35]
  (iii) obtaining access to some part of the callable capital of the World Bank for anti-desertification lending;
  (iv) alternative to (i), (ii) or (iii) above, the earmarking of a modest part of the IMF's existing gold stock as collateral to enable the

independent corporation or other entity to undertake the necessary borrowing.

The remainder of this part will examine in further detail the implications of assumptions (d) and (e) of the previous paragraph with a view to arriving at a range of illustrative magnitudes for concessional lending from IMF-sourced interest subsidy amounts. The approach can be extended without difficulty to accommodate other subsidy sources.

A   Resources mobilizable through combined subsidy-guarantee/collateral mechanisms

The discussion in the feasibility studies above as regards potential interest subsidy amounts arising out of the implementation of a link and of Trust Fund reflows may be summarized as in Table 2. Table 3 (page 120) summarizes the leverage factors with which these subsidy amounts have to be multiplied in order to generate amounts that can be lent out on World Bank Third Window terms. These factors vary according to the market rate at which funds are borrowed and according to whether the loan is fully disbursed in the first year or spread out over a seven-year period. Disbursement in the initial year will approximate to the terms of programme-type loans while disbursements over seven years would be more appropriate to the average anti-desertification project. The Table indicates the leverage factor of 4 which applied in the case of the Bank's Third Window when market interest rates were 8 per cent, and when disbursements were spread equally over the seven-year grace period. Since interest rates are currently at record levels, the yield on World Bank dollar bonds being 15 per cent, it seems reasonable to expect a downward trend for the forthcoming decade. For illustrative purposes it seems convenient to work with an interest rate of 10 per cent and leverage factor of 3 so as to look after the case where disbursements fall between programme and project type lending. [Of course interest rates are no longer at these levels and show an overall decline in all money markets.]

Combining this factor with the subsidy amounts available as illustrated in Table 2 yields a total borrowing potential annually of 240 million SDR during 1982 to 85 and 315 million SDR during 1986 to 1989. These represent magnitudes rounded to the nearest $100 million corresponding to $300 million and $400 million respectively. What remains to be specified are the guarantee collateral arrangements that will permit the borrowing of these amounts to take place.

B Alternative guarantee/collateral arrangements

Ad hoc guarantee arrangements

At least three types of guarantee arrangements are open to governments to establish among themselves for the purpose of underwriting any flotations of bonds raised in the capital markets, irrespective of whether they are to underpin any permanent new development agency.

The first and simplest type of guarantee would be a 'full joint and several guarantee' of the bonds of the participating governments. Each government would, under such a guarantee, be liable for the full amount of the guaranteed obligations and a bondholder could make a claim against any single guarantor for the full amount due. The guarantor governments could, by arrangement determine the proportion of their respective liabilities as among themselves, and any guarantor paying a bondholder more than its agreed share could recover from the other guarantors.

Secondly, and at the other extreme, would be a 'several guarantee' under which each government would guarantee only a specified proportion of each bond. This would present greater difficulty for bondholders since, unless special arrangements were made, it would be necessary in the event of a default to make a claim against each guarantor government.

A third and intermediate possibility would be an arrangement under which each member government, while giving only a several guarantee, would contribute, in an amount based on the proportion of the bonds guaranteed by it, to a fund in which the bondholders would share if there were a default. This approach is illustrated by the Austrian Government Guaranteed Loan (1923–43) arranged by the Financial Committee of the League of Nations. The net effect of such an arrangement would have many of the characteristics of, but would fall short of, a full joint and several guarantee. It is this third approach which may be described as a 'limited joint and several guarantee'.

This type of guarantee would be somewhat similar to the arrangement by which the member Governments of any development agency 'guarantee' that agency's obligations on the basis of its callable capital. Strictly speaking, bonds issued by the agency are not guaranteed by the agency's member governments. However, in case of a default on its bonds, or to prevent a default, the agency could call for payment of the callable as distinct from paid-in capital of its member governments and use the amounts received to make the required payments on its bonds. Each government's obligation to make payments on such calls is not dependent on payment being made by other member governments and, since successive calls may be made until sufficient funds are available to pay the obligations of the Agency, the system is very like a joint and several guarantee. But since each government is liable only to the extent of its uncalled capital, such a system is not equivalent to a full joint and several guarantee, as no single government can be liable for the total amount being guaranteed by all participating governments together.

It follows from this account of possible guarantee mechanisms that any desired system of guarantees can be established among participating governments, whether or not a formal agency is set up. Indeed in the case of anti-desertification, it is open to interested governments to establish a system of 'limited joint and several guarantees' along the lines of the Austrian Loan guarantee procedure referred to above. This would enable them to raise moneys without the formal establishment of an institution in a manner which would obviate the need in the transitional years for the formal putting in of capital subscriptions other than the amounts required for the interest subsidy element. This would, in principle, enable interested governments to put together only the cash amounts required for the interest subsidy element, not only leaving any chosen interested government or designated entity to discharge the task of disbursing funds but also enabling it to float bonds in the capital market on the strength solely of the guarantees being provided by the group of interested governments taken together.

Incorporation in an anti-desertification institution/agency of a callable capital component

The anti-desertification institution described later in this study (the independent financial corporation) would have total equity capital of $100 million all of which would be expected to be paid in. If the institution were to be equipped to borrow for lending on Third Window terms, it would be necessary to supplement this capital with a callable component of $2.8 billion so as to permit market borrowings averaging 300 million dollars in the first four years of operation and 400 million dollars in the subsequent four years. In round figures it would seem safe to work with a callable capital provision of $3 billion.

In this event the implementation of phase 1 could conceivably involve a more gradual stepping up towards this callable capital provision. In the first instance, interested donors from developed and OPEC countries may be approached to put up the callable capital for the first four years amounting to $1.2 billion. An amount of this order could be raised in capital markets through a system of *ad hoc* limited joint and several guarantees if donor countries seem unwilling to make the more formal commitment of converting guarantees into callable capital at the initial stages.

Alternatively, the anti-desertification institution could in principle utilize the callable capital of the World Bank which already has a high credit rating. The negotiations for securing access to this guarantee capability could presumably be conducted by the institutions on terms that would safeguard its autonomy, taking into account the precedents already established between the World Bank and other development institutions, e.g. the Caribbean Development Bank. With the increase in the capital base of the World Bank to $80 billion and the proposed doubling of its gearing ratio, the World Bank's total guarantee capability would rise to $160 billion. Assuming that 10 per cent of this, or $16 billion, would be used for anti-desertification purposes, this would more than suffice to accommodate the lending capability of $3 billion generated by subsidy amounts stemming from IMF sources alone (Trust Fund reflows and SDR-link moneys) for an eight-year programme of development. Indeed, this guarantee capability, if assumed to be spread over eight years at an annual rate of $2 billion a year, would permit additional subsidy amounts to be used. Thus, resources stemming from IMF sources would permit the borrowing of up to $400 million annually. Resources arising out of international trade taxes of $200 million annually would permit the borrowing under guarantee of up to thrice this amount, of $600 million annually. The total to be raised therefore in capital markets under World Bank guarantee of up to $1 billion would still leave a substantial margin of unused guarantee capability to which recourse could be made if necessary by finding other interest subsidy sources, e.g. by varying slightly the rate of taxes on international trade.

Collateral arrangement

The alternatives to 1 and 2 above are, as mentioned, borrowings against collateral which is most easily available in the form of some part of the IMF's existing gold stock. This currently stands at 100 million ounces and can be valued at $400 an ounce or a total of $40 billion. The recommendation that the IMF's gold constitute collateral for borrowing for desirable purposes has been made by the Brandt Commission[36] on the grounds, largely, that not a great deal is to be gained by dissipating the IMF's gold over time if the gold sales procedure were, for example, to be revived. In this particular case of the collateral required to support borrowings based on IMF sources alone, e.g. Trust Fund reflows and SDR-link moneys, the amount of $3 billion compared to the total IMF gold stock of $40 billion amounts to a modest percentage of 7.5 or 7.5 million ounces.

V FEASIBILITY STUDY AND WORKING PLAN FOR THE ESTABLISHMENT OF AN INDEPENDENT FINANCIAL CORPORATION FOR THE FINANCING OF DESERTIFICATION PROJECTS

The proposed independent corporation for the financing of desertification projects is dealt with in chapter V of Part II. In that chapter, attention was drawn to the fact that the establishment of such a corporation 'to attract investments and provide financing for suitable anti-desertification projects with non-commercial rates of return' had been proposed previously. The proposal called for providing the corporation with equity funds from countries with international account surpluses and from financial institutions. It was stated that the feasibility of establishing such a public international corporation for desertification control projects financing would depend on whether donor countries and organizations would provide the necessary resources for its establishment.

In that chapter, the warning was given that the projects the corporation would be expected to finance would for the most part be incapable of bearing interest costs even on the highly concessional terms presently available from IDA and similar financing institutions. It was important to recognize, therefore, that such projects would have to be financed primarily with funds provided on an interest free basis.

The feasibility study undertaken in this study accordingly proceeds on the assumption that, in the view of the General Assembly, a corporation could be provided with funds free of interest.

The same chapter considered the question of whether the corporation should be established as an affiliate or subsidiary of an existing finance institution, or as an independent institution; however, neither method of establishment was recommended above the other. The present study interprets the General Assembly's use of the term 'independent' to describe the corporation in its resolution 35/73 as intended to steer further examination of this issue away from the possibility that this institution might be established as an affiliate or subsidiary of any existing body. The corporation must be independent in the sense of standing on its own and possessing its own resources of equity, loan capital, or both. This does not mean that the finance corporation must be precluded from recourse to any existing institution for technical or administrative co-operation, but rather that any such co-operation would be recognized as extended by one independent institution to another, and not by an institution to its associate or affiliate.

A Equity of the corporation

The first question which must be examined is that of the corporation. Size and composition of equity normally determine both the character and capacity of a financial institution. It is understood, however, that this corporation would use borrowed funds, not its equity, for lending, so the size of its equity is not a matter of the highest importance. All participants in the activities to be financed through the corporation – both donors and potential clients – should be members of the corporation and should contribute to its equity capital.

It has the advantage of simplicity to assign to each member of the corporation the same or equal rights of membership. The equity capital of the corporation could be divided into shares valued at approximately $US 1 million each; acquisition of one share would then be made the qualification for membership in the corporation. Countries and international organizations only would be eligible to purchase shares.

If the large majority of the States members of the United Nations should become shareholders of the corporation, its total equity capital would be of the order of $150 million. A conservative estimate of the participation which is likely to be forthcoming is, however, that a maximum capital of $100 million would be sufficient to provide for all the potential participants.

It is estimated that about $1 million would be small enough to be subscribed in one lump sum by each participating country. There should be no need for deposits and successive calls in the subscription of a sum of this size. It would be appropriate, therefore, to require each member of the corporation to pay for its share of equity in a single instalment; as a result, the equity of the corporation would be fully subscribed in the first year of its operation.

All shares should be equal in status as regard both dividend and repayment of capital.

No member should be able to obtain loan assistance from the corporation until his qualifying share capital has been taken up.

B Research fund

Countries in the donor group should be encouraged, upon subscribing to their qualifying shares, to make further voluntary contributions of from $1 million to $4 million each towards a research fund.

Four-fifths (80 per cent) of the income earned on the equity of the corporation, minus operating expenses, should be credited annually to the research fund until a target of approximately $40 million (from contributions and earnings) is reached.

C Operating expenses

The management of the corporation should ensure that its annual operating expenses would not exceed the amount earned on the subscribed equity and unused loan capital. The corporation should not use its equity to make loans, at least not during the first 10 years of its existence. Operating expenses should be financed from earnings on capital and capital should remain available to meet any unexpected losses which the corporation might incur.

D Management of the corporation

It is proposed that the day-to-day business of the corporation should be managed by a board of not fewer than seven nor more than 11 directors. Each director should have an alternate. The member and the alternate would be selected from different countries within the same group. Each should serve for three years. The following groups are suggested:

Developing countries

Mediterranean Group	Two members
	Two alternates
Latin American and Caribbean Group	One member
	One alternate
Sudano-Sahelian and Tropical African Group	Two members
	Two alternates
Asian and Pacific Group	Two members
	Two alternates

Developed countries

West European and Others Group	Three members
	Three alternates
East European Group	One member
	One alternate

Countries would be assigned to one or another of the above groups upon applying for membership in the corporation.

The board of management should have the usual responsibility and powers to establish internal rules and regulations, to appoint the chief executive and senior staff and to select advisers, including legal advisers.

E Demand for loan funds

The study included in Part II of this book estimated the area of desertified land in developing countries which will require external financial assistance. These figures in millions of hectares were:

	million hectares
Irrigated land	16.35
Rangeland	1,445.24
Rain-fed cropland	97.19
Total:	1,558.78

At the medium estimate of cost, the total cost of a programme of corrective measures on these lands would be as follows:

	$ US million
Irrigated land	12,262.50
Rangeland	18,065.50
Rain-fed cropland	17,007.50
To this should be added:	
Sand dune fixation	449.00
Total:	47,784.50

This represents an average cost of $2,389.22 million ($2.389 billion) per annum over the next 20 years if a minimum period of 20 years is allowed for the meaningful attack on desertification problems which is contemplated in the Plan of Action.

It must be assumed that a larger portion of the expenditure for desertification control will be provided by national governments. (The projects presented to the Consultative Group for Desertification Control up to August 1981 indicate that external aid will probably be required for not more than 72 per cent of the total expenditure.)

On the assumption that the amount required to be financed by loan would remain a major portion of the total expenditure, it may be assumed that the demand for the high concession loans could be accommodated within an annual supply of the order of $500 million per annum for 20 years.

F Estimate of potential supply

Ever since the General Assembly at its Seventh Special Session considered the supply of resources for global development, the proposal has been actively advanced and supported that the resources to be provided in the future should be made available on an increasingly assured and predictable basis. This means in practice that those who use the resources should be able to plan their use with reasonably accurate prior knowledge of the amounts and terms on which the resources will become available over the years. This proposal was in itself a substantial modification of, and withdrawal from, the earlier advocacy of automatically flowing resources. Some measure of assurance of supply and predictability must therefore be written into the arrangements for the interest-free loans to the corporation. It seems reasonable that the minimum level of predictability to be provided for should be a seven-year supply of money. Taking account of the time which will be spent in the disbursement of desertification control funds, it seems that the period to be covered by the minimum forward look should be seven years.

Donor countries wishing to contribute towards long-term desertification projects through the corporation should therefore be requested to undertake to supply a specific sum of loan money to the corporation each year over a seven-year period. This annual contribution could be comparatively small.

Desertification is neither a merely national nor a merely regional phenomenon, but a global problem. All the governments of the world should be involved in the proposed loan programme. The progressive annual withdrawal of 20 million hectares of land from production of any kind is a phenomenon the consequences of which will soon be felt over the whole world – in mass movements of population; in the appearance of refugees on a scale with which no nation or group of nations will be able to contend; in malnutrition and food shortages; and in the persistent demand for short-term measures which will absorb all available resources without providing a single long-term solution. It is essential, therefore, that all countries of the globe, whether or not they are directly threatened by the advance of the deserts, should consider themselves involved in the fight against desertification and that those countries directly threatened should be in the vanguard. All member countries of the United Nations should be encouraged to contribute towards the interest free resources of the corporation.

Recognizing that all governments have a duty to participate in the campaign against desertification, it is useful to consider next which countries are best qualified to take the lead. We consider that these countries with the greatest economic strength might be invited to contribute one half of the total resources required, i.e. $250 million per annum, on the basis shown opposite.

The remaining $250 million should be the subject of negotiations with the countries with centrally planned economies, the petroleum-

	Annual provision of interest-free loans in $US millions
United States of America	50
Federal Republic of Germany	36
Japan	30
France	24
United Kingdom	20
Netherlands	18
Italy	12
Australia	10
New Zealand	10
Sweden	10
Norway	10
Denmark	10
Canada	10
	$250

exporting countries and the remainder of the developing countries members of the United Nations (except the least developed). Countries in these groups should contribute towards this total according to their United Nations scale of assessment.

G Availability of loans from the corporation

It is recognized that some countries with arid and semi-arid areas will prefer not to approach the corporation with a request for finance. For the most part they will regard it as sufficient if they receive technical assistance, help with planning, and so on. It cannot, of course, be assumed that no applications for finance will be made to the corporation in respect of lands not currently regarded as arid or semi-arid. All that can be said on this point is that:

(a) Countries will be eligible for financial assistance from the corporation only for anti-desertification projects to be carried out in or for arid or semi-arid areas: that is, areas having less than 50 centimetres of rainfall per annum.
(b) Loans will be available towards the cost of composite projects, but only for that part of the project which qualifies on the basis of anti-desertification criteria, i.e. which involves measures directly designed to arrest the spread of desert conditions.

The following example will illustrate these principles: Country A submits to the corporation a project which involves both the rehabilitation of irrigated land which has become water-logged and salinated and the addition of irrigation to previously non-irrigated lands. The corporation would finance only that part of the project which involved the rehabilitation of the land. Country A would be free to seek assistance from other agencies such as the International Bank for Reconstruction and Development, IFAD, etc., for the part of the project involving addition of irrigation works.

H Funding of research and experimentation

It is envisaged that a great deal of experimentation would characterize the early desertification control projects. Typically, technical assistance missions would visit a country or region, assess anti-desertification needs and recommend that a pilot project be implemented and evaluated before a larger-scale project is attempted. Such pilot projects may involve moving of livestock in order to try new grazing methods, or experimental planting of new crops introduced from other climates. Governments will wish to seek assistance, by way of grants or loans, towards the implementation of such pilot projects. These loans and grants would be made from the corporation's research fund.

I Terms of the loans

Once it is decided that a full-scale desert control project should be undertaken and the project is designed, the national governments, or the governments of the region (where a region is involved), should negotiate with the corporation as to what part of the total cost they will be able to bear. Financing would then be sought for the balance of the cost.

The commitment to make the loan having been accepted by the corporation, it is proposed that a uniform disbursement period of five years should be allowed. Actually, the disbursement period could be a year or more longer or shorter. During this five-year period, no interest would be charged on the funds disbursed and no commitment fees would be imposed.

From the sixth to the twenty-fifth year, interest should be payable at 2.5% per annum on the sums lent, and no repayment of principal should be required. (This grace period may be extended to the thirtieth year if necessary.) No interest should be capitalized.

From the twenty-sixth to the fortieth year the loan should be repaid together with interest at 2.5 per cent per annum on the diminishing balance. (In order to equalize the annual payments, these could take the form of a 15-year annuity with interest at 2.5 per cent per annum.)

It is an essential feature of this financial outline that the interest paid in during the sixth to the twenty-fifth or thirtieth year, together with accumulated earnings thereon, should be sufficient[37] to repay the original interest-free loan from the donor. The original donor-lender should have the loan repaid before the end of the thirtieth year. Receipts by way of principal and interest from the thirty-first year onwards would be available to make further desertification loans.

J Responsibility for the loans

Money borrowed from the corporation will be available (a) to national governments; and (b) with a guarantee from the national governments to domestic entities identified by the national governments.

K Comparison with IDA and IFAD concessionary loans

It will be noted that the terms suggested for high-concession loans by the corporation are somewhat less generous than those currently available on the most highly concessional terms which IDA and IFAD allow. IDA loans are repayable in 50 years, and attract no interest at all; there is only a commitment charge of three-fourths of 1 per cent. IFAD makes loans on 'regular', 'intermediate' and 'highly concessional' terms. The last category – the 'highly concessional' – carries no interest rate. It attracts a service charge of 1 per cent per annum, and has a maturity period of 50 years, including a grace period of 10 years.

According to information available, loans on these highly concessional terms are available only to countries having special characteristics – the least developed countries in the case of IDA, and the 'food priority' countries in the case of IFAD. The least developed countries have less than $350 per capita of GNP. Up to the present time countries can only qualify for inclusion in the IFAD category of food priority countries on the basis of:

(a) low per capita income;
(b) a projected cereal deficit by 1985;
(c) severe protein-calorie malnutrition;
(d) insufficient average increase in food production;
(e) serious balance-of-payments constraints.

It follows that the corporation's terms should be less generous than those of IDA or IFAD in order to encourage countries which qualify for IDA and IFAD assistance for anti-desertification projects to seek such assistance. It is understood that the corporation's high-concession loans should be reserved for projects that would not currently qualify for financing because of the difficulty of estimating in advance potential financial returns.

As the corporation gains experience in the management of its interest-free loans, it will certainly be subjected to increasing pressure to augment the types of loans it can offer. By this time as well it may also have accumulated modest reserves. It will no doubt consider it appropriate then to endeavour to raise additional resources by the issue of its own bonds. It should be possible to issue bonds equivalent in total to the amount of its share capital. Loans from these resources would be made on less generous terms than the loans from interest free resources.

L Project implementation, control and monitoring

At the present time the management and implementation of the PACD is the responsibility of the Governing Council and the Executive Director of the United Nations Environment Programme (UNEP) and the Administrative Committee on Co-ordination (ACC). The General Assembly has, by its resolutions, effected institutional arrangements to assist UNEP to fulfil its mandate. These arrangements include:

(a) the expansion of the mandate of the United Nations Sudano-Sahelian Office (UNSO), as a joint venture between UNEP and UNDP to cover the implementation of the PACD in the Sudano-Sahelian region;
(b) the establishment of the Desertification Unit within the UNEP secretariat;
(c) the establishment of an Inter-Agency Working Group on Desertification;
(d) the establishment of a Consultative Group for Desertification Control (DESCON) to assist in mobilizing resources for the activities undertaken within the framework of implementation of the Plan of Action.

1 United Nations Sudano-Sahelian Office

The original functions of UNSO were to assist the drought-stricken countries of the Sahel in implementing their medium- and long-term rehabilitation and development programmes. Under joint arrangements between UNEP and UNDP, the scope of its concern was extended to become the focal point and central co-ordinating mechanism of the United Nations system for assisting 19 countries of the Sudano-Sahelian region, on behalf of UNEP, in implementing the Plan of Action. To achieve these objectives, the Office provides assistance in planning, programming and implementation of priority projects, and in resource mobilization.

The sources of finance available to UNSO are its own Trust Fund, voluntarily contributed by United Nations members, and the joint contributions of UNEP and UNDP to its operating expenses. The Office canvasses donor agencies bilaterally and multilaterally in an effort to broaden the financial basis for the funding of projects in this region.

2 Desertification Unit

This Unit was established in 1978. It co-ordinates and follows up activities relating to the implementation of the PACD. The Unit serves as the secretariat office for the Inter-Agency Group on Desertification and the Consultative Group for Desertification Control. [In 1985 it became the Desertification Control Programme Activity Centre.]

3 Inter-Agency Working Group on Desertification

The Group was established in 1977. It includes representatives of the relevant members of the United Nations system. It compiles and disseminates information, monitors projects and makes recommendations regarding constraints and priorities for implementation of the Plan of Action for consideration by the ACC and then by the Governing Council of UNEP.

4 Consultative Group for Desertification Control

The Consultative Group (DESCON) is composed of co-sponsors and other members. Co-sponsors include some members of the United Nations system. Other members are certain States having an interest in, or suffering from, desertification such as the United States, Mexico, Iran, Iraq, etc., and some other members of the United Nations system and some financial institutions, for example, the Arab Bank for Economic Development in Africa. It considers project proposals submitted through the Sudano-Sahelian Office, the Desertification Unit, or other United Nations organs, examines the need for additional financing and assists in the mobilization of resources.

It should not be assumed that, with the setting up of the corporation, any of these agencies will be rendered superfluous. A country in the Sahel which desires to undertake a desertification control project will still need the advice and help of the Sudano-Sahelian Office. A country outside the Sahel region which has not yet embarked on its share of the Plan of Action will need the advice of the Desertification Unit so as to assess its needs and establish its priorities. When a project has been designed, the government of the country concerned would approach potential suppliers of credit and would determine the eligibility of the project, or parts thereof, for IDA, IFAD, the World Bank or Development Bank finance.

It will be sufficient to assume that the management of the corporation would join the other agencies in the Inter-Agency Working Group and the DESCON and participate in the scrutiny of all projects.

M Procedure: establishment of the corporation

If the proposals in this study are accepted by the international community, the group proposes that the charter of the new corporation should be embodied in an agreement in the form which follows.

The agreement should take effect and the corporation should come into being when 30 countries, including 10 countries, donors of interest-free loans, have signed or adhered thereto.

AGREEMENT

The Governments on whose behalf this Agreement is signed:

Recognizing the importance of the Plan of Action to Combat Desertification;

Realizing that concerted action must be taken to promote the flow of resources on an assured and predictable basis into the financing of long-term desertification control projects; and

Bearing in mind the necessity to provide a substantial portion of such long-term funds on an interest free basis;

Have agreed to the following Charter of the International Finance Corporation for Anti-desertification;

CHAPTER I – GENERAL PROVISIONS

Article 1

Establishment of the Corporation

The International Finance Corporation for Anti-desertification (hereinafter called the Corporation) is hereby established, and shall function in accordance with the provisions of this Agreement.

Article 2

Legal Status

The Corporation shall possess an international legal personality and shall enjoy administrative and financial independence.

Article 3

Headquarters

The Headquarters of the Corporation shall be Kenya. Such agencies, branches or representative office of the Corporation as the Board of Directors may consider necessary, may be established in other countries.

Article 4

Objectives and Functions

The objective of the Corporation is to provide finance for measures to control desertification in the territories of member countries and so accelerate the economic and social development of these countries.

In fulfilling this objective, the Corporation shall be authorized and shall have the power:

(a) to raise funds of grants and loans from its members;
(b) to raise funds by the issue of bonds and securities;
(c) to issue and recover loans to member countries, or to institutions and agencies identified by member countries, for the control of desertification either alone or jointly with any other institution; and
(d) to provide technical assistance and services in the field of anti-desertification.

Article 5

Membership

1 The countries which have signed this Agreement shall be the original founding members of the Organization;

2 Any other country of the United Nations or international financial institution or inter-governmental agency, may accede to membership in the Corporation upon a favourable decision taken by a simple majority vote of the Board of Directors of the Corporation;

3 The liability of members in the Corporation, as shareholders, shall be limited to the amount, if any, which may remain unpaid on the portion of capital subscribed.

CHAPTER II – FINANCIAL RESOURCES

Article 6

Resources of the Corporation

The resources of the Corporation shall consist of:

(i) the equity capital subscribed by its members;
(ii) voluntary subscriptions to its capital by members and other parties;
(iii) interest-free loans to the Corporation, amounts recovered through the repayment of loans and interest received on loans;

(iv) other amounts raised by borrowing; and
(v) funds derived from the operations or otherwise accruing to the Corporation.

Article 7
Equity Capital

The initial capital of the Corporation shall be $US 150 million in shares of $US 1 million each. Each member of the Corporation shall be required to subscribe one share. Shares subscribed shall not be transferable except in accordance with the provisions of this Agreement.

Article 8
Alteration of Capital

1 The Corporation may increase or decrease its initial share capital, may cancel, consolidate or divide shares or otherwise alter its capital by a resolution of a general meeting;
2 No member shall be obliged to subscribe additional amounts in the case of general or individual increases in the capital of the Corporation.

Article 9
Additional Subscriptions

The Board of Directors shall lay down appropriate conditions and procedures for additional subscriptions to be provided by members on a voluntary basis. Such additional subscriptions may include contributions to a Research Fund.

Article 10
Loan Capital

1 The Corporation may accept interest-free loans from members or non-members and may repay such loans from its resources.
2 When it is in a position to do so, the Corporation may also raise funds by the issue of bonds or other securities, or by obtaining credit in national or international capital markets.
3 Members making interest-free loans to the Corporation shall be required to make such loans annually for periods of seven years, or such further periods as may be agreed between the members and the Board.

Article 11
Periodic Review of Resources

The Board of Directors shall between the fifth and the seventh year, and at intervals of not more than seven years thereafter, review the resources of the Corporation in relation to its needs and may submit recommendations thereon to a meeting of the general membership.

CHAPTER III – OPERATIONS

Article 12

Principles

The Corporation shall provide assistance in the form of grants or loans for desertification control measures (including measures on irrigated land, rangeland, rain-fed cropland and for the fixing of sand dunes), to be carried out in or for arid or semi-arid areas – that is, areas having less than 50 cm of rainfall per annum.

The Corporation may complement its available resources of finance in the best interest of its members, but shall not provide highly concessional loans to any undertaking which in the opinion of the Board is able to obtain adequate financing elsewhere on terms which the Corporation considers reasonable.

Article 13

Eligibility for Loans

Subject to the conditions stipulated in this Agreement the recipients eligible for assistance from the Corporation shall be:

(a) Governments of member countries as well as agencies, departments or subdivisions thereof;
(b) public or private institutions or enterprises operating in the territory of member countries;
(c) interregional or subregional organizations established by member countries for the promotion of anti-desertification measures.

Article 14

Types of Loans

The Corporation may issue as the Board of Directors may deem appropriate:

(a) highly concessional loans – being loans for periods exceeding 30 years with low interest and a grace period of not less than 5 years;
(b) less concessional loans – being loans for a period not exceeding 30 years, with a higher rate of interest and a shorter grace period.

Article 15

Terms and Conditions of Loans

General guidelines for each type of loan shall be issued by the Board of Directors.

Article 16

Limitation on Financial Operations

When this Corporation proposes to raise funds by the issue of bonds or other securities or by obtaining credit in capital markets, the Board of Directors shall ensure that the total amount outstanding of loans, credits and guarantees issued by the Corporation (other than interest-free loans) shall not at any time exceed twice the amount of its subscribed capital, its reserves and free income.

CHAPTER IV – ORGANIZATION AND MANAGEMENT

Article 17

Structure of the Corporation

The Corporation shall have a general meeting, a Board of Directors, a President and such other officers and staff as may be required for the performance of its duties.

Article 18

The General Membership

(a) Each member of the Corporation shall appoint a representative and an alternate representative to attend meetings of the general membership.

(b) Each representative shall have one vote in the general meetings of the Corporation. An alternative representative may vote only in the absence of the representative.

(c) General meetings shall be held not less frequently than every two years.

(d) The general meeting is the supreme authority of the Corporation and shall have the power to:

  (i) issue or amend general policy guidelines for the management of the Corporation's business;
  (ii) consider and approve the annual report by the Board of Directors, or the President, on the Corporation's activities;
  (iii) elect the Board of Directors;
  (iv) appoint auditors and set their remuneration; and
  (v) transact any other business not within the competence of any other organ.

(e) Unless otherwise stipulated, resolutions put to the vote in a general meeting shall be decided by a simple majority of members. Resolutions adopted in a general meeting shall be binding on all members.

Article 19

Board of Directors

1 The Board of Directors shall be composed of no fewer than seven nor more than eleven directors. Each director should have an alternate, the director and the alternate to be selected from different countries within the same group of members. Each director and alternate director should serve for three years. An alternate director shall not vote in a meeting of the Board, except in the absence of the director.

2 Directors and alternate directors shall be selected according to groups:

Developing Countries

The Mediterranean Group	2 directors 2 alternates
The Latin American and Caribbean Group	1 director 1 alternate
Sudano-Sahelian and Tropical African Group	2 directors 2 alternates
Asian and Pacific Group	2 directors 2 alternates

Developed Countries

Western European and Other States Group	3 directors 3 alternates
East European Group	1 director 1 alternate

3 Countries shall be assigned to one or another of the above groups upon applying for membership in the Corporation;

4 Each director and alternate shall serve until a successor has been appointed.

Article 20

Powers of the Board of Directors

All the executive powers of the Corporation shall be vested in the Board expent such powers as are reserved to the general meeting. These powers shall include the power:

(a) to formulate policies in accordance with the provisions of this Agreement and such guidelines as the general meeting may issue from time to time;
(b) to adopt regulations and other measures to ensure the efficient operation of the Corporation;
(c) to decide on borrowing, the issue of bonds and securities or guarantees;
(d) to prepare for meetings of the general membership;
(e) to appoint the President and senior staff of the Corporation
(f) to approve the Corporation's budget; and
(g) to interpret the provisions of this Agreement.

Article 21

Resolutions of the Board

Resolutions at meetings of the Board shall be decided by a majority of the directors, each director having the right to cast one vote only. The President shall have the casting vote in the event of an equal division of votes.

Article 22

The President

The President of the Corporation shall be appointed by the Secretary-General of the United Nations upon the recommendation of the Executive Director of the United Nations Environment Programme for a term of five years, which may be renewed for another term of equal duration. He shall remain in office until his successor assumes office.

Article 23

Other Staff

In selecting its personnel, the Corporation shall, subject to the paramount importance of securing a high standard of efficiency and technical competence, pay due regard to the importance of recruiting personnel on as broad a geographical basis as possible.

Article 24

International Status of Staff

1 In the discharge of their duties, members of staff owe their duties entirely to the Corporation and to no other authority. They shall refrain from any act which is incompatible with the international character of their functions and independence.

2 Each member of the Corporation shall respect the international status of the staff and shall refrain from all atempts to influence any member of staff in the discharge of his duties.

CHAPTER V – FINANCIAL PROVISIONS

Article 25

The Financial Year

The financial year shall commence on 1 January and end on 31 December of each year. The term of the first financial year shall be fixed by the Board of Directors.

Article 26

The Budget

The President shall submit to the Board of Directors on a date not later than 15 October of each year an estimate of the expenses and current income of the Corporation for the following financial year.

Article 27

Accounts and Auditors

1 The President shall ensure that proper books of account are kept which

will give a true and fair view of the state of the Corporation's affairs and will explain its transactions.

2 The Board of Directors shall submit to the general meeting of the Corporation a report containing an audited statement of the accounts including a balance sheet and an income and expenditure account. The form of such statement shall be determined by the Board of Directors.

3 The accounts of the Corporation shall be certified by a firm of Auditors of recognized (International) standing appointed by the general meeting.

Article 28

Profits and Reserves

The general meeting of the Corporation shall determine, upon the recommendation of the Board of Directors, the allocation which should be made of the net income of the Corporation.

CHAPTER VI – IMMUNITIES AND PRIVILEGES

Article 29

Immunities of Assets, Correspondence and Records

1 The property and other assets of the Corporation in the territories of member countries shall enjoy immunity from nationalization, confiscation or any form of seizure by executive or legislative action. Such immunities shall not extend to judicial action or to assets purchased by proceeds of loans extended by the Corporation to its beneficiaries.

2 The official correspondence and records of the Corporation shall be accorded in each member country the same privileges enjoyed by the official communications and records of other member countries.

Article 30

Exchange Restrictions

The assets and transactions of the Corporation shall not be subject to the exchange control regulations prevailing in any member country.

Article 31

Immunity from Taxation

The Corporation, its assets, property, income and its operations and transactions authorized by this Agreement as well as the shares of its capital shall be immune from all taxation and all customs duties in member countries.

Article 32

Personnel Privileges and Immunities

All directors, alternates, officers and employees of the Corporation, including experts on missions for the Corporation, shall be immune from legal process with respect to acts performed by them in their official capacity.

CHAPTER VII – SUSPENSION OF OPERATIONS AND LIQUIDATION

Article 33

Temporary Suspension

In an emergency the Board of Directors may temporarily suspend its activities in respect of new operations pending an opportunity for further consideration and action by the general meeting.

Article 34

Liquidation

1 The general meeting may, by a two-thirds majority and after giving member states not less than three months' notice, decide to terminate operations and liquidate the Corporation.

2 The Corporation shall undertake the liquidation proceedings either by itself or through a Committee of liquidators to be appointed by the general meeting.

3 No distribution of assets shall be made to members on account of their subscriptions to the capital of the Corporation until all liabilities to creditors have been discharged or provided for.

CHAPTER VIII – MISCELLANEOUS PROVISIONS

Article 35

Arbitration and Interpretation

1 If a disagreement should arise between the Corporation and a country which has ceased to be a member or between the Corporation and a member after a decision to terminate operations of the Corporation, such disagreement shall be submitted to arbitration by a tribunal of three arbitrators. One arbitrator shall be appointed by the party instituting the proceedings, another by the adverse party, and the third, within 30 days of the appointment of the second arbitrator, by the President of the International Court of Justice.

2 A majority vote of the arbitrators shall be sufficient to reach a decision which shall be final and binding upon the parties.

3 Questions of interpretation or application of this Agreement shall be submitted to the Board of Directors for decision.

Article 36

Amendments to the Charter

This Agreement may be amended by a resolution passed by a three-fourths majority at a general meeting of the Corporation. Only member countries or the Board of Directors shall be authorized to propose amendments to this Agreement. Amendments shall enter into force for all members three months after date of the approval.

Article 37

Prohibition of Political Activity

Neither the Corporation nor any personnel working in any of its organizations shall in any manner interfere in the political affairs of any member country or in other international political issues.

Article 38

Relations with other Organizations

The Board of Directors may conclude with organizations of national, regional or international character, agreements conducive to the expansion of co-operation with the Corporation.

CHAPTER IX – FINAL PROVISIONS

Article 39

Entry Into Force

This Agreement shall enter into force when approved by resolution of the General Assembly.

Article 40

First Meeting of the Board of Directors

The Governing Council of the United Nations Environment Programme or the Executive Director of the Programme shall convene the first general meeting of the Corporation within a period of six months following the entry into force of this Agreement.

Article 41

Commencement of Operations

The Board of Directors shall inform all member countries of the date of commencement of operations.

VI   CONCLUSIONS

A Additional means of financing

Summing up the prospective yields of the various additional measures examined in the preceding feasibility studies, this study presents the following amounts and time frames:

(a) *General trade taxation*: Assuming a uniform 0.1 per cent rate, revenues are projected to rise from $2.0 billion in 1980 to $2.5 billion by 1984 and $3.0 billion by 1987, etc., for an indefinite period into the future. The relatively high degree of predictability of these flows would permit the organization charged with their collection to pledge them as security behind borrowing at market rates, if deemed advisable, with the possibility of subsidizing the costs of such borrowings to the extent that they are devoted to lend for designated purposes, such as desertification control measures. Sources of funds for such interest subsidies have been spelt out in Section A of Chapter III earlier.
(b) *IMF gold*: Reflows to the IMF Trust Fund from book profits from earlier sales of gold not otherwise earmarked would amount to $228 million (SDR 250 million) in each of the calendar years 1986 to 1989, subject to vote by the Fund's Executive Board; additionally, collateralization of 7.5 per cent of the IMF gold holdings would support borrowing of perhaps $1.5 billion to $2 billion spread over several years. Profits from future sales of IMF gold depend on future prices and are, therefore, inherently unpredictable.
(c) *Proceeds of the proposed indirect SDR-link*: These would yield SDR 800 million ($920 million), in each of the years from 1983 to 1987, assuming IMF agreement on a fourth basic period, (possibly followed by a fifth period from 1988–92), provided that a sufficient number of developed countries accept this type of link.
(d) *Parking fees for communication satellites*: These are estimated at $100 million per year beginning around the year 1990, but subject to considerable uncertainty regarding both amount and timing. Earlier receipts, but of substantially lesser amounts, as shown on page 145, are a possibility.
(e) *Revenues from deep sea-bed mining*: In the absence of agreement on the present draft Treaty, the expectations of levels of future revenue generated under these or similar conditions remain too vague to permit meaningful quantification at this time. The present draft Treaty designates the International Sea-bed Authority as the sole body entitled to receive any net revenue from sea-bed mining. If the current negotiations do not result in a convention, the possibility of obtaining revenues for international purposes from sea-bed resources should be pursued separately.
(f) *Common Fund, Second Account*: Designed to improve the production and marketing of specific commodities, other than price stabilization, the Second Account could become a source of funds for anti-desertification measures whenever the authorities designated to speak for a particular commodity felt such support to be justified. Pending such determination, even though the Treaty has entered

into force, no estimate can be made of such possible flows of funds.

Altogether, and subject to the margin of error implicit in these projections, the following estimates summarize the prospective annual flows from 1982–90, assuming, in each case, timely activation of the requisite measure.

B Modalities of obtaining resources on a concessionary basis

Resources can be obtained from governments on a concessionary basis as a form of official developmental assistance, or they can be obtained from governmental or private sources on commercial terms and converted to concessional resources by interest subsidies. Every dollar of interest subsidy can create several times as much in the way of concessionary resources. The exact rates will depend on the terms of the commercial borrowing and the degree of concessionality desired.

Providing the lender with reassurance that the loan will be repaid is crucial. This can be done by offering collateral or some form of guarantee, or possibly both.

Collateral could be money raised in any number of ways, or physical assets, such as gold, pledged by international financial institutions or governments. One possibility to be explored is the use of part of the gold of the IMF as collateral.

Participating governments could provide a joint and several guarantee under which each government would assume responsibility for repayment of the entire amount of a given loan, or they could provide merely a several guarantee under which the liability of each government would be limited to a specified fraction of the loan. An intermediate arrangement would be for participating governments to contribute to a fund on which creditors could draw in the event of a default.

This last arrangement is somewhat similar to the more common use of callable capital as backing for a loan. The callable capital could in principle be that of the borrowing institution or that of an international financial institution. In the latter case, the Financial Institution would guarantee repayment of the loan and might also act as the agent of the borrowing institution in obtaining it.

The strength and credibility of the reassurances provided to lenders – in the form of collateral or guarantees or both – would determine whether borrowing was possible and on what terms. The terms of a loan would in turn determine the size of the interest subsidies required to achieve a desired degree of concessionality. The adequacy of reassurance to lenders is therefore the essential component of any effort to mobilize resources on concessional terms.

C Independent financial corporation

The General Assembly's request that the group of experts should provide a feasibility study and working plan for a financial corporation implies a willingness on the part of the international community to supply such a corporation with interest-free resources. The experts had little difficulty, therefore, in designing an independent operating financial corporation for the financing of desertification projects.

The plan calls for a corporation independent of other financial institutions, with a minimum capital of $100 million subscribed by its members in equal shares of $1 million each. It is intended to operate with a predictable flow of interest-free loans contributed by the majority of the member countries of the United Nations. The contributors to these 24-year loans would be asked to pledge a supply of loan funds in the first instance amounting to $500 million per annum for seven years. They would be requested to replenish the supply after intervals of seven years.

It is important to recognize that the corporation's *raison d'être* is to operate on the extreme end of the project feasibility scale. It will finance those long-term desertification projects which, although possessing definable economic and social viability, will produce returns which cannot be quantified at the time of project preparation, and so the project would not be entertained by the existing financial institutions. About one-fifth of all desertification expenditures is expected to fall into this category, representing some $500 million out of an annual total expenditure of $2.4 billion.

The ultimate borrower of interest-free money would be offered loans for periods of between 40 and 45 years. Nothing would be paid on the loan in the first five years but a nominal payment of 2.5 per cent per annum would be required over the remainder of its life. The payments over the first 20 years would be regarded as interest, after which the remaining payments would retire the loan in 15 to 20 years.

As the corporation gains experience in the management of its loan portfolio, the possibility of raising additional resources through the issue of its own bonds would be considered.

D Co-ordinated financing plans

Two broad sources of funds to finance the PACD have been described above: additional means of financing and interest-free loans to an independent financial corporation. Only a fraction (possibly 10 to 15 per cent) of funds raised by additional means of financing could, however, be expected to be available for anti-desertification purposes.

Table 2 on page 118 presents estimates of the total amounts that might be obtained by the various additional means of financing over the period 1982 to 1990. With one exception, the funds raised by additional means of financing would be turned over initially to a new central institution or treasury established for that purpose. That institution would make allocations for specific purposes, including anti-desertification.

Money provided by the Second Window of the Common Fund for Commodities would presumably be earmarked from the beginning for a specific anti-desertification activity and would not be deposited in the treasury.

Funds raised by additional means of financing and made available for anti-desertification purposes could be administered by the existing Environment Fund, the independent financial corporation, or the Special Account for Desertification established in 1978.

Money raised by additional means of financing could be used directly for anti-desertification purposes, or they could be used as interest subsidies to permit the re-lending on concessional terms of funds borrowed on commercial terms. This kind of borrowing and re-lending could be undertaken by the independent financial corporation.

Although the additional means of financing and the independent financial corporation can be usefully related, they are not interdependent. Both approaches to financing the PACD can and should be pursued separately.

The corporation should be created as soon as possible with the capital and assets described in earlier sections. Its assets could be expanded later by drawing on funds raised by additional means of financing. At some point, a decision could also be made on whether the corporation should borrow additional funds using its augmented assets as collateral. Guarantees by interested governments might also be necessary. In time, the corporation might arrange for callable capital as a substitute for guarantees.

The estimated yield of additional means of financing plus the initial assets of the independent financial corporation could support an average annual anti-desertification effort of about $400 million over the period 1982 to 1990.[38] This is less than one-fourth of the average annual financing of $1.8 billion additional external funds that is required to carry out the Plan of Action over the next 20 years. Even though annual expenditures in the next few years need not be as high as the average for the 20-year period, additional financial resources are clearly needed.

Such resources might be obtained by increasing the yield of the additional means of financing (for example, the trade tax rate might be raised above the assumed 0.1 per cent), by allocating more than 10 per cent of the funds raised by additional means of financing to anti-desertification, or by borrowing additional money on commercial terms and using prospectively available funds as interest subsidies to permit re-lending such money on concessional terms.

Notes

1. See the report of the Symposium on the Interrelations between Resources, Environment, People and Development, held at Stockholm, from 6 to 10 August 1979, United Nations publication, Sales No. E.80.II.A.8.
2. Includes some expenditure in areas not currently threatened with desertification.
3. Plan of Action to Combat Desertification (PACD) adopted by the Conference on 9 September 1977 and approved by the General Assembly in resolution 32/172 of 19 December 1977.
4. Eleanor B. Steinberg and Joseph A. Yager with Gerard M. Brannon, *New Means of Financing International Needs* (Washington, DC, Brookings Institution, 1978).
5. *North-South: A Programme for Survival*, Report of the Independent Commission on International Development Issues under the Chairmanship of Willy Brandt (Cambridge, MA., MIT Press, 1980), pp. 290–91.
6. Requirements of a comprehensive system of international financial co-operation (TD/B/C.3/161 and TD/B/C.3/161/Supp. 2).
7. *New Means of Financing International Needs*, pp. 54–55 and appendix A.
8. Greater precision regarding elasticities – at least in the long run – would offer useful clues regarding the incidence of the burden on importers and exporters. Unfortunately, the great variety of goods and commodities involved does not permit such precision.
9. However, exemptions could be granted for trade between England and Scotland, Belgium and Luxembourg, and the United States and Puerto Rico, which does not formally clear customs.
10. At historic exchange rate current value is approximately $1.15 per SDR (24 July 1981).
11. See IMF Press Release 1980/81 of 19 December 1980 from which the following account is drawn.
12. While the Extended Fund Facility took the first step in reducing the distance between the short-term balance-of-payments approach of the Fund and the long-term approach of the Bank, there remains a substantial gap between the two approaches involving the medium-term. 'It has become clear in the course of the '70s that for developing countries the process of adaptation to external shocks to the balance-of-payments of the magnitude experienced during this period calls for efforts of adjustment that go well beyond the time frame and scope envisaged under the Extended Fund Facility. No doubt as the period of adjustment is lengthened and the scope of measures required expand, it becomes

difficult to distinguish between the adjustment process and the development process. The two activities reach a point at which they may be said to merge. The problems of adjustment and development are, therefore, a continuum and the stratification of institutions should not prevent related problems from being dealt with in a related manner.' (See United Nations Balance of Payments Adjustment (1979) Report to the Group of 24, pp. 12–13.)
13 Lester B. Pearson *et al.*, *Partners in Development* (1969), p. 225, World Bank publication, Washington, DC, USA.
14 The most recent report to the Group of 24 of the UNDP/UNCTAD Project (March 1981) on measures to strengthen the SDR proposed that during the next basic period the annual allocations of developing countries other than those in structural surplus should be increased to 150 per cent of their quota share and the allocations to industrial countries decreased by an equivalent amount and that this should be brought about by a voluntary action on the part of industrial countries. This is tantamount to an indirect link somewhat differently formulated than we have above with virtually the same effect in terms of link transfer amounts.
15 See document ID/IPC/CF/CONF/24.
16 Paragraph 3 (a) of article 18 ('the second account') of the Agreement states: 'The measures shall be commodity development measures, aimed at improving the structural conditions in markets and at enhancing the long-term competitiveness and prospects of particular commodities. Such measures shall include research and development, productivity improvements, marketing and measures designed to assist, as a rule by means of joint financing or through technical assistance, vertical diversification, whether undertaken alone, as in the case of perishable commodities and other commodities whose problems cannot be adequately solved by stocking, or in addition to and in support of stocking activities.'
17 The OPEC Fund has agreed to pay for the subscriptions of certain low-income developing countries.
18 *North-South: A Programme for Survival*, p. 245.
19 United Nations, *Treaty Series*, Vol. 499, p. 311.
20 Under the same paragraph 2 of article 173, such net revenue would be permitted to fund certain of the capital needs of the Enterprise, and/or to compensate developing States (but not commercial enterprises) for losses of their market share from existing land-based mining operations, if the Convention's production limitations should prove insufficient.
21 Richard Cooper, 'The Oceans as a Source of Revenue', in J. Baghwati (ed.) *The New International Economic Order: The North-South Debate* (Cambridge, MA, MIT Press, 1977), p. 112.
22 *New Means of Financing International Needs*, p. 149; and D. Leipziger and J. Mudge, *Seabed Mineral Resources: The Economic Interests of Developing Countries* (Cambridge, MA, Ballinger, 1976), p. 148.
23 *A Cost Model of Deep Ocean Mining and Associated Regulatory Issues* (J. D. Nyhart, Lance Antrim, Arthur E. Capstaff, Alison D. Kohler, Dale Leshaw), MIT, 1 March 1978; Research Institute for International Techno-Economic Co-operation of Technical University Aachen and

Battelle-Institut e. V. Franfurt, *Analysis of the MIT Study on Deep Ocean Mining – Critical Remarks on Technologies and Cost Estimates* (Franz Diederich, Wolfgang Müller, Wolfgang Schneider), March 1979; Eleanor B. Steinberg and Joseph A. Yager with Gerard M. Brannon, *New Means of Financing International Needs* (Washington, DC, Brookings Institution, 1978); Danny M. Leipziger and James L. Mudge, *Mineral Resources: the Economic Interest of the Developing Countries* (Cambridge, MA, Ballinger, 1976); United Nations Committee on Natural Resources, Report of the Secretary-General on 'Mineral Resources: Trends and Salient Issues, with Particular Reference to Molybdenum, Cobalt and Vanadium, including Problems of Technology Transfer' (E/C.7/115); Arthur D. Little, Inc., *Technological and Economic Assessment of Manganese Nodule Mining and Processing*, Revised, Nov. 1979 (Cambridge, MA), prepared for United States Department of the Interior, Office of Minerals Policy and Research Analysis.

24 *Third United Nations Conference on the Law of the Sea, Official Records*, vol. XII, (document A/CONF. 62/C.1/L.26, Annex E, Table 1).
25 *North-South: A Programme for Survival*, p. 245.
26 *Sea-bed Mineral Resource Development: Recent Activities of the International Consortia*, document ST/ESA/107.
27 Deep Sea-bed Hard Mineral Resources Act of 1980, House of Representatives bill 2759–33.
28 See, for example, *New Means of Financing International Needs* (Washington, DC, Brookings Institution, 1978), pp. 27–28.
29 *North-South: A Programme for Survival*, p. 245.
30 Thus, the INTELSAT V satellites now on order by the International Telecommunications Satellite Organization use polarization and beam separation to achieve multiple use of both the 14/11 GHz frequency band and the 6/4 GHz band used by older satellites. The new satellites have a capacity of 12,000 telephone circuits, compared with only 4,000 to 6,000 in earlier versions (INTELSAT, *Annual Report*, 1 April 1979–31 March 1981, p. 10).
31 INTELSAT, *Annual Report*, 1979–81, p. 9.
32 Ibid., p. 21.
33 World Bank, *World Development Report 1980*, p. 29.
34 As events turned out, only $154 million of subsidy amounts were raised from developed and OPEC countries, permitting the raising of no more than $700 million worth of capital.
35 It would be understood that (i) could always phase into (ii) at the discretion of the governments concerned.
36 *North-South: A Programme for Survival*, p. 211.
37 Compounded and accumulated interest at 8 per cent per annum will permit full repayment of the principal in 19 years.
38 This estimate assumes that 10 per cent of the $33.66 billion from additional means of financing would be allocated to anti-desertification and that $200 million of the $500 million in interest-free loans received by the corporation would have been re-lent by 1990.

List of participants in group of high-level specialists in international financing who prepared the study contained in Part III

(Serving in Personal Capacities)

Sir Egerton Richardson (Chairman)
Former Financial Secretary of Jamaica;
Permanent Representative of Jamaica to the United Nations

Roberto Campos
Ambassador Extraordinary and Plenipotentiary of Brazil to the Court of St James;
Former Minister for Planning and Coordination, Government of Brazil

Ingemund Bengtsson
Former Minister of Labour,
Speaker of the Parliament of Sweden

A. A. Hegazy
Former Prime Minister and Minister of Finance,
Government of Egypt

Paul-Marc Henry
Former Assistant Administrator, UNDP, and President, Development Centre, OECD

Driss El-Jazairy
Ambassador Extraordinary and Plenipotentiary of Algeria to Belgium, Luxembourg and the European Economic Community

M. Kassas
Professor of Plant Ecology,
University of Cairo, Egypt

Mansour Khalid
Former Foreign Minister,
Government of the Sudan

Lal Jayawardena
Ambassador Extraordinary and Plenipotentiary of Sri Lanka to Belgium, Netherlands, Luxembourg and the European Community;
Former Secretary, Ministry of Finance and Planning, Sri Lanka

Paul Gerin-Lajoie
President, Projecto International Inc., Montreal, Quebec, Canada;
Former President, Canadian International Development Agency

Philip Ndegwa
Chairman, Kenya Commercial Bank; Former Deputy Assistant Executive Director (Programme), United Nations Environment Programme

Keichii Oshima
Professor of Nuclear Engineering, University of Tokyo;
Former Director for Science Technology and Industry (OECD)

Adolph J. Warner
Former Vice-President and International Economist,
Salomon Brothers (Investment Bank)

Joseph Yager
Senior Fellow,
Foreign Policy Studies,
The Brookings Institution

CONVENOR
Mostafa K. Tolba
Executive Director, UNEP

SECRETARIAT
Yusuf J. Ahmad
Acting Assistant Executive Director,
Desertification Branch, UNEP

Anastase Diamantidis
Deputy to the Director,
Regional Office for Europe, UNEP

Index

Africa
    areas of desertification 4, 63–4, 78
    crisis xvi, 106
    reclamation programme 5–6
    Sahara 4, 5, 64
aid *see* assistance
air travel, taxing 36, 109, 110
Americas
    areas of desertification 63, 77
    domestic fiscal capacity 79
    see also United States of America
Antarctica 40–1
aquifers 6
arms
    expenditures, reduction 47–8
    sales taxation 8, 16, 45–6
Asia 6, 65, 78, 79
assistance, external 8–11, 20, 32
    assessed budgets 33
    division, 1978 74–5
    injection 82
    level 12, 28, 32
    need for 28, 74–5, 157
    official development assistance 9–10, 20, 74–5, 106
    related to areas of desertification 66
    voluntary 32
Australia 65, 71
Austrian Government Guaranteed Loan 151, 152

banks, development xii, 29, 82, 86, 87
Brandt Commission
    North-South: A Programme for Survival 27, 33, 35, 36, 42, 43, 46, 50, 52, 55, 108, 122, 141, 153
Brookings Institution
    New Means of Financing International Needs xi, 108, 109, 113–14, 115

Capital Development Fund 19
capital markets *see* world capital markets

coffee producers 129, 130, 131
commercial schemes 81–2
Common Fund (CF) xv, 17, 30, 31, 52–3, 57, 126–32
    current status 128–32
    First Account 127
    Second Account 127–8, 174–5
    Second Window 31, 57, 177
'Common Heritage Fund' 39, 43
communications tax 110
concessionary resources, obtaining xii–xiii, 147–53, 175–6
Consultative Group on International Agricultural Research (CGIAR) 93–6
corporation, public international, proposed ix, 29, 86–90, 154–78
    control 162–3
    equity 155
    establishment 164
        articles 164–73
    feasibility 87–8
    loans 166, 167–8
        availability 159–60
        compared with IDA and IFAD 161–2
        demand for funds 157
        proposed annual, by country 159
        responsibility 161
        terms 160–1
    management 157, 168–70
    operating expenses 156
    research fund 155–6, 160
    resources 165–6
    supply estimate 157–9
corporations, multi-national 54
cotton producers 130, 132
currency, local 82

Desert Encroachment Control Rehabilitation Programme (DECARP) 6

desertification
  areas viii, xvii, 59–60, 157
    related to assistance required 66
  causes 4
  control programmes xi–xii
    see also Plan of Action to Combat Desertification
  countries affected by assistance from 9, 20
  effects viii–ix, 19, 158
  increase, rate viii, xvii, 4, 27, 59–60
  limitation 73–4
  magnitude of problem viii, 59–60, 105
  monitoring 6, 72–3
  premises about 2
developed countries
  assistance from see assistance
  desertification in 27
  domestic resources 76
developing nations xi–xii, 12, 27–9
  domestic resources 28, 76–83
development assistance see assistance, external
disarmament 15–16, 45–8, 58
domestic resources 28
  fiscal capacity 77–80
  mobilization 76–83

Economic and Social Council 18, 38
energy crisis 72
environmental programmes xiii
equity investment 11, 18
Europe
  desertification areas 65
  foundations 90, 91, 93, 94, 100
  land loss other than by desertification 98
Exclusive Economic Zones (EEZ) 12–13, 39, 43–5, 57, 133–4
experimentation 160
  see also research

finance see Plan of Action to Combat Desertification
financing institutions, national development xii, 82–3, 85
  see also banks
fiscal capacity of affected countries 77–80

fisheries 38–9, 41
fodder 72
food production 7
  'food for workers' 81–2
  World Food Programme 28
foundations
  grants 29–30
  guides to 99–100
  largest, by country 94–5
  research and training 11, 90–6
funding see Plan of Action to Combat Desertification
funds-in-trust 19

Germany, Federal Republic
  foundations 90, 93, 95
  legislation 140
gold see International Monetary Fund
goods, marketed 82
government loans 20, 82, 147–53
  concessionary 10–11, 28–9, 83–5, 147–8, 175–6
grants 28, 29–30, 85–6, 89, 90–2, 147
green belts 5–6, 70
guayule 81

International Bank for Reconstruction and Development (IBRD) 19
International Commodity Agreements (ICA) 53, 57, 126–32
International Commodity Body (ICB) 127–8
international commons 12–14, 31, 38–42
International Development Association see World Bank
International Development Fund 51–2
International Finance Corporation for Anti-Desertification see corporation, public international, proposed
International Fund for Agricultural Development (IFAD) 19, 89, 161
International Labour Compensatory Facility (ICLF) 36–7
International Monetary Fund (IMF)
  concessionary loans 147, 148
  Extended Fund Facility 120–1, 178–9
  gold sales 30, 51–2, 55, 56, 118, 121–2, 148, 174

gold stock 56, 153
special drawing rights (SDR) 15, 30–1, 48–50, 55, 57, 118–26, 148–50, 153, 174
link with development finance 122–6
Trust Fund 118–22, 124, 148, 174
international needs xi, 12
international organizations, budgets 33
see also names of organizations
International Sea-Bed Authority 43–4, 133–40, 174
International Telecommunications Satellite Organization (INTELSAT) 14, 41, 141, 144, 146, 180
International Telecommunications Union (ITU) 141–2
irrigated land xiii, 27, 59, 60, 62, 63–6, 68, 73, 98, 157

Japan 98
jojoba 81

labour 28
International Labour Compensatory Facility (ILCF) 36–7
land
aridity 4, 27, 59
irrigated xiii, 27, 59, 60, 62, 63–6, 68, 73, 98, 157
loss, other than desertification 61, 98
rainfed cropland 27, 59, 60, 63–6, 68, 69–70, 73, 158
rangeland 27, 59, 60, 62–9, 71–2, 73, 98, 158
see also desertification
Least Developed Countries (LDC) 83
livelihood systems 71–2
livestock 71–2
loans, concessionary 10–11, 28–9, 82–6, 147–53, 157, 175–6
estimated annual, by country 159
lottery, world-wide 53

manganese nodules 12–13, 31, 40, 57, 134, 136, 139
Massachusetts Institute of Technology (MIT) 139

military expenditure and taxes 15–16, 45–6, 58
mineral resources *see* ocean resources; oil
moon 40, 141

New Means of Financing International Needs see Brookings Institution
North-South: A Programme for Survival see Brandt Commission
Norway 98

ocean
pollution 13–14, 42
resources xv, 12–13, 31, 34, 38–40, 132–40, 174
living 38–9
magnitudes of revenue 139–40
oil
pollution 13–14, 42
tax 8, 35–6
Organization for Economic Co-operation and Development (OECD), 74, 85
development assistance 9–10, 20, 74–5
provided, 1978 106
foundations 90, 91–2
Organization of Petroleum Exporting Countries (OPEC) xiii, 74, 85, 88, 106, 129

Pearson Commission 123
Plan of Action to Combat Desertification (PACD) viii–ix, xi–xvi, 2–3, 5, 26–7
allocation of funds 55–6
cost viii, xii, 6–7, 19–20, 27, 28, 62, 67–8, 69–70, 71, 75, 106–7
breakdown 73
financing ix, xii, xiii, 19
automaticity ix, xiv, xv, 7–8, 18, 20, 32–58, 107
burden-sharing between countries 17, 54–5
co-ordinated plans 176–8
financial plan 58–75
scope and objectives 60–1
funds mobilization and management 17–19, 54
means inventory 32–58
sources 8–17, 28–32
specially established 12–17, 30–2

Plan of Action to Combat
   Desertification – *cont.*
   guidelines 76–7
   implementation, xiii, 2–3, 19, 162–4
   needs 7, 8, 106–7
   premises 2
   projects 5–6
   studies
      1978 xiii–xiv, xix, 1–21
         participants preparing 22–3
      1980 xiv–xv, xix, 25–100, 104–5
         participants preparing 100–2
      1981 xv–xvi, xix, 103–80
         participants preparing 181
   see also corporation, public
      international, proposed
pollution, taxing 13–14, 42
population viii, 4
programmes, international 12

regional projects 77
research 29–30, 90–6
   funding 155–6, 160

Sahara 4, 5, 64
sand dunes 60, 61, 68, 70–1, 73
satellites
   parking charges 14, 31, 41–2, 57–8, 140–7, 174
      financial considerations 144–6
      overcrowding prospects 143–4
      taxation principles 142–3
Saudi Arabia
   tax proposals 8, 36, 45
sea *see* Exclusive Economic Zones; ocean; United Nations Conference on the Law of the Sea
Soviet Union *see* Union of Soviet Socialist Republics
Spain 65
special drawing rights (SDR) *see* International Monetary Fund
Sudan 6, 64
   Sudano-Sahelian region 5, 6, 11, 64, 78, 79, 80, 82

taxation
   arms sales 8, 16, 45–6
   consumption 8, 37–8
   domestic 80–1
   international xiv–xv, 8, 14–15, 35–6, 54–5, 111–18
   of invisibles 36, 109–10
   military 15–16, 45–8, 58
   oil 8, 35–6
   of pollution 13–14, 42
   revenue 35–6
   of technology transfer 36–7
   telecommunications 14, 41–2
   of transport 36, 109, 110
   see also ocean resources; satellites, parking; trade
technology transfer, taxing 36–7
telecommunications
   taxing 14, 41–2
Third World *see* developing nations
trade
   commodities stabilization 52–3
   flows 14–15, 31, 35–6, 37, 57
   imports as GNP percentage, countries' 112–14
   taxing 108–18, 174
      commodities 31, 35–6
      invisible 36
      surpluses 37
      world 14–15, 31, 35–6, 37, 54, 57, 108–10, 111
training 29–30, 90–6
transport, taxing 36, 109, 110

Union of Soviet Socialist Republics (USSR)
   areas of desertification 65, 71
   disarmament 48
   proposal, natural resources 44
United Nations
   assistance from 74, 106
   Committee for Development Planning 8
   'Commodity Trade and Economic Development' report 33–4
   Conference on Desertification (UNCOD), viii, ix, xi, xii, xvi, 2–6
      see also Plan of Action to Combat Desertification
   Conference on the Exploration and Peaceful Uses of Outer Space 147

Conference on the Law of the Sea 12, 13, 31, 38–9, 41, 43–4, 55, 57, 133, 134, 139–40
Conference on Science and Technology 37
Conference on Trade and Development (UNCTAD) 17, 49, 50, 53, 109, 111
Consultative Group for Desertification Control (DESCON) xvi, 55, 162, 163
Development Programme (UNDP) 19, 89, 162
Economic and Social Council (ECOSOC) 18
Committee for Development Planning 38
Environment Programme (UNEP) viii, xvii
Desertification Unit 74, 162, 163
fund xii
Special Account 32, 35
PACD implementation 55, 93, 162–3
PACD studies xiii, xix, 26, 47, 54, 86, 99
General Assembly
resolutions
disarmament 46, 47–8, 97
military budgets 16, 47–8
ocean resources 39
official development assistance 9–10
outer space 141
Plan of Action to Combat Desertification xii, 3, 17, 19, 33
technology transfer 37
and studies xiv, xv–xvi, 26, 104
Group of Governmental Experts on the Relationship between Disarmament and Development 46, 48
Group of 24 49, 50, 179
Group of 77 37, 49–50
Institute for Training and Research (UNITAR) 96
Inter-Agency Working Group on Desertification 162, 163
Sudano-Sahelian Office (UNSO) xix, 32, 93, 99, 162–3
University (UNU) 96
see also Plan of Action to Combat Desertification
United States of America 44
areas of desertification 63, 71
disarmament 48
foundations 90, 91, 92–3, 94, 99–100
legislation 140
proposal, natural resources 44

Vienna Programme of Action of Science and Technology for Development 37

wheat producers 129, 130
World Bank xii, 29, 84, 86, 88, 120–1, 153
International Development Association xii, xiii, 19, 50, 84, 86, 87, 88, 161
Third Window 148, 149, 150, 152
world capital markets xii–xiii, 10–11, 20, 28, 85–6
concessionary loans 147–53
World Development Fund 56
World Food Programme 28
World Health Organization (WHO) 33
'World Solidarity Contribution' 38